AF386732

Fuzzy Management Methods

With today's information overload, it has become increasingly difficult to analyze the huge amounts of data and to generate appropriate management decisions. Furthermore, the data are often imprecise and will include both quantitative and qualitative elements. For these reasons, it is important to extend traditional decision making processes by adding intuitive reasoning, human subjectivity and imprecision. To deal with uncertainty, vagueness, and imprecision, Lotfi Zadeh introduced fuzzy sets and fuzzy logic. In this book series "Fuzzy Management Methods", fuzzy logic is applied to extend portfolio analysis, scoring methods, customer relationship management, performance measurement, web reputation, web analytics and controlling, community marketing and other business domains to improve managerial decisions. Thus, fuzzy logic can be seen as a management method where appropriate concepts, software tools and languages build a powerful instrument for analyzing and controlling the business.

Hans-Heinrich Bothe • Edy Portmann

Computing with Words

An Introduction to Fuzzy Logic Use-Cases with Theory and Applications

 Springer

Hans-Heinrich Bothe
Institute for Electronic Engineering
Technical University of Berlin
Berlin, Germany

Edy Portmann (iD)
Human-IST Institute
University of Fribourg
Fribourg, Fribourg, Switzerland

ISSN 2196-4130 ISSN 2196-4149 (electronic)
Fuzzy Management Methods
ISBN 978-3-032-24116-0 ISBN 978-3-032-24117-7 (eBook)
https://doi.org/10.1007/978-3-032-24117-7

This work is derived from Fuzzy Logic: Einführung in Theorie und Anwendungen, by Hans-Heinrich Bothe, 978-3-540-56967-1, Springer, 1995 and represents an updated and expanded translation of the original German edition.

This Springer imprint is published by the registered company Springer Nature Switzerland AG
The registered company address is: Gewerbestrasse 11, 6330 Cham, Switzerland

If disposing of this product, please recycle the paper.

fuzzy logic flows
infinite truth values dance
beyond black and white

This poem by Andreas Meier (2026) is a
Japanese haiku.
It promotes the search for truth through
multi-values.

Special Note

My father, Hans-Heinrich Bothe, co-author of this book, was a dedicated researcher and a teacher to many. He was driven by deep curiosity and a genuine enthusiasm for research, paired with a desire to inspire others. His excitement for the things that fascinated him was truly contagious. When I was still a child, he died in a tragic mountain accident, but the enthusiasm he carried has remained with me to this day.

Throughout his career, he worked in various fields and countries. But he took the greatest pride in his role as a researcher and professor, and in the impact he was able to make through it. Aside from his expertise in fuzzy logic, many of his projects were devoted to supporting people with disabilities, especially those with hearing impairments.

I am grateful to Edy Portmann, who shares my father's humanistic approach to engineering, for honoring his work and for translating and extending this textbook. It is my hope that, as you read and explore the concepts presented in this edition, you will also sense the spirit behind my father's work: a belief that knowledge can have meaningful applications in helping people.

Cologne, Germany   Henrike Witt
Spring 2026

Preface

Humanist engineers are individuals who not only possess the technical skills of engineers but also incorporate a humanist perspective into their work. This means they consider the ethical, social, and environmental implications of their projects, aiming to create solutions that benefit humanity and the planet. Based on a German book by first author Hans-Heinrich Bothe (1995), which was kindly provided by his daughter, Henrike Witt, as a basis for the translation, this textbook was written with a humanist engineering mindset.

Melanie Fletcher, Patan Imran Khan, Lars Hubacher, Rebecca Evans, and Todd Beanlands supported us in the translation and book development, and Irene Soraya Gavilanes contributed us her graphics. Barbara Bethke, Christian Rauscher, Jialin Yan, and Sneha Arunagiri supported us in the production of the book. Finally, next to all other unnamed helpers, we owe thanks to the father of fuzzy logic, the humanist engineer of the first hour, Lotfi Zadeh. Humanist engineers create holistic solutions to challenges by tailoring technology to human needs and interests. We employ a broad range of research methods steeped in fuzzy mathematics and science. Humanist engineers as we, investigate the interaction of people with technology as well as technical development.

Students of a humanistic engineering graduate with a foundation in designing natural user experiences and interfaces, creating clear information visualizations, conducting usability research, designing for the web, and building technologies. In doing so, their models often draw on people's spoken language as an interface. And as its name suggests, computing with words is a humanistic engineering methodology in which language is used rather than numbers for computing reasons. Thereby fuzzy logic plays a key role, which is why Zadeh (1996), equated it with computing with words. It is vital, when available information is too imprecise to justify the use of exact numbers, or, when there is an acceptance of approximation and imprecision, which is exploited to achieve tractability, robustness, low solution cost, and better fit with our human realities. Exploiting this tolerance is of utmost importance in fuzzy logic-based computing with words.

In this sense, a word is viewed as a label of a granule, that is, a fuzzy set of points drawn together by similarity, with fuzzy sets playing the role of fuzzy constraints on a variable. The premises are implicitly expressed as propositions in a natural language. In coming years, computing with words is expected to evolve into a basic methodology with wide-ranging ramifications on both basic and applied levels. In this textbook, we take a closer look at fuzzy logic-based computing with words methods. The book is designed for humanist engineers looking for a forthright introduction to fuzzy logic technology and computing with words applications.

Fribourg, Switzerland   Edy Portmann
Spring 2026

References

Bothe, H.-H. (1995). *Bewertung mit unscharfen Mengen*. Technische Hochschule Ilmenau.

Meier, A. (2026). *A crate of haikus: Art book*. Pako Publisher.

Zadeh, L. A. (1996, May). Fuzzy logic = Computing with word. *IEEE Transactions on Fuzzy Systems, 4*(2).

Competing Interests The authors have no competing interests to declare that are relevant to the content of this manuscript.

Contents

List of Figures

Chapter 1
Introduction

Abstract This chapter introduces fuzzy logic. This logic is a computing approach based on degrees of truth rather than the strict true/false (1 or 0) of binary logic, allowing systems to model human-like reasoning, ambiguity, and uncertainty. It uses ranges of values to define partial truths.

Fuzzy logic is an extension of binary logic (e.g., see Bothe, 1995). Its mathematical and historical background is provided by work on multivalued logics (e.g., L3 logic according to Lukasiewicz, 1932), which arose in particular in connection with the uncertainty of events in quantum theory. In these works, the truth values true and false (or 1 and 0) of a statement, which are possible in classical binary logic, were supplemented by further intermediate states (e.g., undefined or ½).

Zadeh (1965) extended this theory to provide a mathematically exact description of variables with linguistically and thus fuzzily defined values. The rules for linking these variables form the axiomatic framework for fuzzy logic. In the years that followed, a large number of theoretical and application-oriented works were produced, particularly in the fields of:

- Control and regulation
- Process monitoring and diagnostics
- Pattern recognition
- Medicine and psychology
- Business and economics
- Computer science and engineering
- Logic and mathematics

The primary aim with fuzzy logic is not to open up areas of application whose specific problems cannot be solved by other methods. If the use of fuzzy logic methods leads to innovative products, it is rather because product development is simplified and development time is significantly reduced.

While Łukasiewicz logic is based on numerical truth values from the unit interval [0, 1], fuzzy logic describes truth values on a linguistic value scale. The values are expressed in verbal terms such as *very false, false, true, and very true* and

H.-H. Bothe, E. Portmann, *Computing with Words*, Fuzzy Management Methods, https://doi.org/10.1007/978-3-032-24117-7_1

are mapped to numerical truth values using characteristic functions. The decisive step toward applying this logic theory in the field of technology was to allow general variables with linguistic values and map them to the corresponding physical value scales. This opens up the possibility of transforming verbal expressions into a mathematically comprehensible domain where they can be automatically processed.

Two classic tasks in system analysis and development are to understand the physical relationships involved and to formulate them using mathematical methods. In real systems, it is often particularly difficult to establish sufficiently accurate system equations, whereas the physical relationships can be easily described in words. This also applies to the determination of the necessary controller behavior for stabilizing complex processes or meaningful rules for classifying complex objects, if such fuzzy expert knowledge is available. When fuzzy methods are used, the task of formulating the physical relationships shifts to using well-defined words to describe the problem and implementing them quantitatively.

Computer-aided engineering methods convert the verbally formulated rules for system description either with the aid of a preprocessor into a high-level language code or directly into the machine code of microprocessors. Both conventional microprocessors and specially developed fuzzy processors can be used. The code can be integrated into the existing programs, thus expanding the range of options for variable types and operations.

The use of fuzzy logic in consumer electronics also makes it interesting to develop integrated components which perform a large part of the necessary operations directly at the circuit level (e.g., Heite et al., 1989, Watanabe et al., 1990, and Yamakawa & Miki, 1986). In view of the fact that digital electronics has become established, this process is leading to the development of fuzzy electronics, which will share its areas of application with digital electronics. In this sense, fuzzy logic also represents an extension of binary logic. Work has also been underway since the mid-1980s on an application in the field of artificial intelligence . A combination with the field of neural networks (e.g., Kosko, 1992 and Surmann et al., 1992) is particularly promising.

1.1   The Concept of Fuzziness

The term "fuzzy" can be interpreted as vague or imprecise . The graded evaluation of the truth value of a statement or piece of information can be regarded as an essential characteristic of the "fuzziness" of a statement or piece of information. With the help of fuzzy logic, descriptions based on (fuzzy) colloquial and technical statements (rules) can therefore also be processed.

Example 1.1 From a medical point of view, clinical severity III in high blood pressure can be described as follows: "Systolic blood pressure higher than 200 Torr or diastolic blood pressure between 120 and 130 Torr. Blood pressure level mostly stable. Very often signs of cardiac insufficiency, changes in the back of the eye, more severe symptoms, partial cerebral disorders" (cf., Bocklisch, 1988).

The terms written in italics indicate a lack of clarity in the respective partial statements, which allows for a certain degree of flexibility in decision-making (in this case, the doctor's diagnosis) and thus enables intelligent behavior. This notation will be retained in the following.

The evaluation of individual statements is carried out from a specific perspective, which may change, also in conjunction with the environmental situation: Significant influencing parameters are often unknown or were not taken into account when the rules were established. This inevitably leads to "uncertainties," especially in complex processes. However, the combination of the existing vague individual statements about the system, which are—perhaps erroneously—assumed to be invariant, has a high degree of truth value and represents special expert knowledge on the basis of which decisions can be made.

Since the prerequisites of probability theory do not necessarily have to be fulfilled when compiling the rules, for example, because the influence of certain parameters has not been recorded, there is a significant difference between uncertainty and the probability of a fuzzy statement. For the sake of simplicity, it should be noted at this point that uncertainty can also be interpreted as a "possibility." A detailed discussion of the terms is provided in Chap. 5. For a better understanding, readers should have mastered the basic knowledge presented in Chaps. 2 and 4.

Various types of uncertainty can be specified: Noise causes stochastic uncertainty (e.g., when evaluating a measured input signal), a lack of information causes informal uncertainty (e.g., when there is insufficient knowledge about the reaction behavior of a technical process), and linguistic formulations cause lexical uncertainty (e.g., in the field of mechanics, *high frequency* means something different than in the field of electrodynamics). System analysis and influence based on fuzzy methods take into account uncertainties in the measured values as well as in their transmission and processing. To explain the term "uncertainty" in more detail, the terms "system" and "model formation" should therefore be introduced.

System
A system is understood to be parts of observable or measurable reality that can be captured using a mathematical description method. Systems can usually be subdivided into subsystems that are interconnected and connected to the environment and exchange information via communication channels. Systems are also components of higher-level metasystems.

Example 1.2 From a medical point of view, the heart can be regarded as an independent system. Anatomically, it consists of several subsystems, such as the heart valves, heart chambers, various muscles, and nerve fibers. These consist of a large number of individual cells, which in turn can be regarded as independent systems. The heart is also part of the higher-level cardiovascular system. It is involved in a complex exchange of information with all the important organs of the body, which is controlled on the one hand by chemical substances such as hormones and on the other hand by electrical currents, which are essentially responsible for nerve conduction.

The system behavior can be influenced by certain input variables as coupling variables to other systems or be memory-dependent and thus time-variant. This time variance often conceals signs of aging, wear, or unaccounted influencing variables, which can lead to problems in technical systems. For example, when developing electronic circuits, the thermal resistance of the components does not have to be taken into account if their expected operating temperature is below a certain limit value. In extreme situations, however, such conditions can easily be exceeded, which—if the influencing variable "temperature" is not taken into account—can even cause a system failure.

Depending on the conditions and perspective, a number of significant and insignificant coupling variables arise; it is up to the developer to identify these and take them into account accordingly.

The sequence of states in a system over time is referred to as a process. In process control, higher-level decisions often have to be made based on a large amount of information that must be weighted. In many cases, these decisions have to be made by humans (e.g., operators, experts), because automatic evaluation based on conventional methods fails. The art of the expert then lies in making the right decisions even with imperfect or vague information (about input variables or the system).

Model Formation
Even when defining the object of interest as an independent system and the resulting coupling variables with the real environment, compromises are made that lead to a more or less accurate (precise) model of the system and the selection of the coupling variables to be taken into account; this creates a difference between simulation and actual behavior.

However, simulation results often contain a relatively high, or at least usable, degree of truth. This raises the question of how the fuzzy truth content of the existing statements can be made usable (e.g., for process control).

1.2 Key Concepts of Fuzzy Logic

The term "fuzzy logic" is used today in both, a narrow and a broad sense.

Narrow Interpretation of Fuzzy Logic
The narrow interpretation describes a logical system in the mathematical sense, with the aim of establishing models for the manifestations of human reasoning and decision-making. These manifestations are defined in a symbolic form in an approximate-rather-than-exact manner.

Broad Interpretation of Fuzzy Logic
The broad interpretation describes the theory and applications of fuzzy sets (i.e., the study of sets with fuzzy boundaries). The significance of this theory lies in the fact that:

- An extraordinary number of natural sets are fuzzy rather than sharply defined.
- Via Boolean algebra , the laws of classical set theory can be transferred to classical logic and switching algebra, and the laws of fuzzy set theory can be derived from them.

Fuzzy Sets
A fuzzy set $\mathbf{M}$ is defined using a characteristic function μ_M, which is referred to as a "membership function" (Chap. 2). In the normalized case, the values of μ_M lie between 0 ($\hookrightarrow$ no membership) and 1 ($\hookrightarrow$ full membership). In contrast to classical set theory, this allows "fluid" transitions of element membership to a set (or the applicability of a statement) to be described. This offers the possibility of "elastic" modeling within the framework of idealizations in system description. The case of sharply defined sets is included in the case of fuzzy sets in that the membership functions can only take the values 0 and 1.

In the following, fuzzy sets will always be presented in bold, while crisp sets will be written in normal font.

Example 1.3 The introduction of fuzzy evaluations will be demonstrated using the two-digit fuzzy relation "much greater than" ($\gg$). In the case of the crisp relation "greater than" ($>$), specifying one value sets a crisp boundary for possible values of the other. However, the restriction in the case ($\gg$) is rather fuzzy; the value range can only be defined inadequately with the help of crisp sets . For clarification, let us consider the fuzzy set $\mathbf{A} = \{x \in X \mid x \gg 1\}$. An attempt at a classical description of the numbers $x \gg 1$, for example, with the help of the set $A^* = \{x \in X \mid x \geq 100\}$, would paradoxically lead to

$$x_1 = 100 \quad \text{is } \textit{much greater } \text{than 1,}$$

$$x_2 = 99 \quad \text{is } \textit{not much greater } \text{than 1.}$$

Example 1.4 The membership function $\mu_M(x)$ shown as a dashed line in Fig. 1.1 describes the set of all real numbers ≥ 18.

For example, it can represent the set of all "legal adults," since the term "legal adulthood" (age of majority) is clearly (crisply) defined by law; however, it is only suitable as a rough, simplified representation of the set of all adult citizens, since no sharp age limit can be specified for the characteristic of adulthood. It would also be unsatisfactory to define all room temperatures that are not uncomfortably cool using a simple step function. A smooth transition—drawn as a continuous line—is much more suitable here; it creates the degree of freedom that, under certain circumstances, a room heated to 16 °C can still be described as not uncomfortably cool. It should also be noted that humidity also influences the assessment of what is considered not uncomfortably cool.

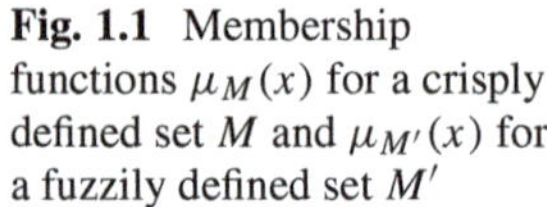

Fig. 1.1 Membership
functions $\mu_M(x)$ for a crisply
defined set M and $\mu_{M'}(x)$ for
a fuzzily defined set M'

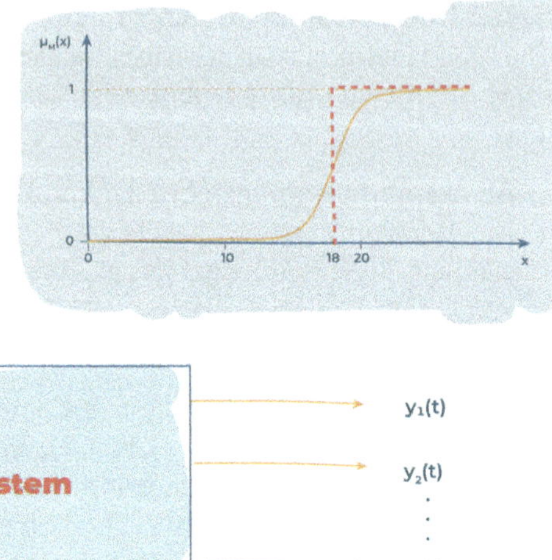

Fig. 1.2 Controller as a fuzzy system with n time-dependent input variables $x_i(t)$ and m time-dependent output variables $y_j(t)$

It should be expressly noted here that a high degree of membership, for example, a temperature value to the set of rooms that are not uncomfortably cool, does not automatically imply a high probability of occurrence of this temperature value. Conversely, a stochastic statement—based on this subjective assessment—may differ significantly from the assessment of a specific expert.

Basic Operating Principle of a Fuzzy Controller

To conclude this chapter, we provide a brief insight into the basic methodology used when applying fuzzy methods. The operating principle of a simple fuzzy controller is particularly helpful for understanding this approach. Figure 1.2 shows the basic block diagram of a controller with n input variables $x_i(t)$ and m output variables $y_j(t)$. In the general case, both the input and output values and the description of the controller behavior are fuzzy.

The behavior of fuzzy controllers is determined by a series of rules that directly convert the input values of the controller into the output values. These rules form a rule base. According to Preuß (1992), a fuzzy controller can also be regarded as a special type of engine map controller . When designing such a controller, therefore, the rules must be defined, on the one hand, and, on the other hand, the question arises as to which algorithm should be used to quantitatively convert the rules into mathematical instructions. The key concepts of "linguistic variables," "fuzzification of input variables," "approximate reasoning," and "defuzzification of output variables" are relevant for this task. They will therefore be explained in more detail below.

Fig. 1.3 Syntactic structure of the linguistic variable "speed" with the primary term *fast*, the antonym *slow*, and a selection of possible modifiers

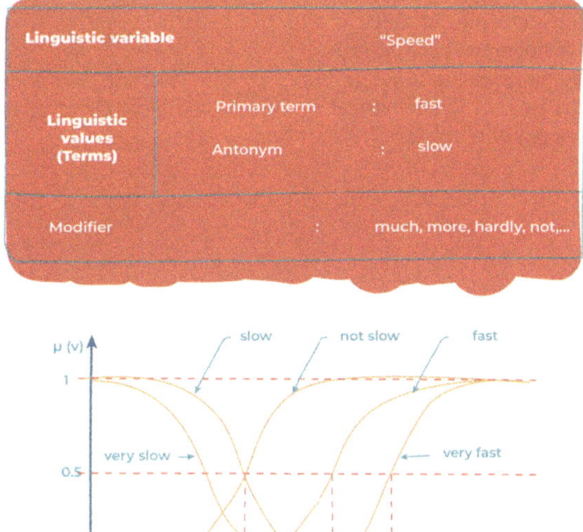

Fig. 1.4 Terms and membership functions of the linguistic variable "speed"

Linguistic Variable

The values of a linguistic variable (also abbreviated as LV) are words or terms of a natural (i.e., synthetic or standardized) language. They are represented by fuzzy sets A_i or their membership functions $\mu_i(x)$ in the form of distribution functions over a base variable x of a (physical) basic domain X. These membership functions map a linguistic scale to a numerical value scale. In addition to creating rules, the developer is responsible for dimensioning this mapping. However, the usually difficult problem of translating physical facts into a mathematical formalism is considerably easier when using fuzzy (linguistic) methods, as software development tools are generally available.

Example 1.5 A possible syntactic structure for the linguistic variable "speed" is shown in Fig. 1.3.

The linguistic values here are initially the primary term *fast* or its antonym *slow*. They can be graded with the modifiers *very*, *more*, *hardly*, *not*, etc. A possible compound value is *very fast*. Logical links such as *very fast or very slow* are also permitted. Possible membership functions (e.g., defined on the basis of expert knowledge) for the individual terms are shown in the following figure. It represents the conversion of a linguistic value scale to a numerical value scale, whereby the physical meaning of the terms is defined (Fig. 1.4).

Fuzzification of the Input Signals

The input values x_i of the controller are crisp numbers. They must first be mapped to the linguistic value scale on which the controller behavior is specified in the form of IF…THEN…rules, since the output values $y_i(x)$ are derived on the basis

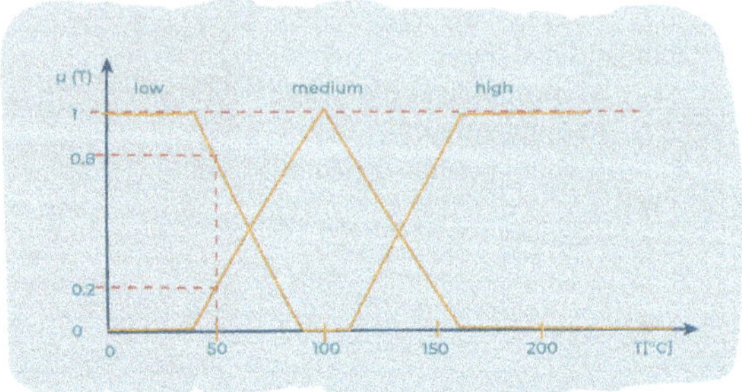

Fig. 1.5 The three terms of the linguistic variable "temperature" are described by membership functions on the basis of the numerical temperature values

of these rules. The x_i are therefore first transformed into the membership space of the linguistic terms involved. If a linguistic variable is described by n terms, the fuzzified signal is an n-dimensional vector $s(x)$ with the elements $\mu_i(x) \in [0, 1]$, $i = 1, \ldots, n$ (cf. Example 1.6). This is also referred to as the "vector of sympathy" of the input value. This vector is further processed when deriving the conclusions. The transformation is called "fuzzification of the input signals."

Example 1.6 (Fuzzification) The temperature of a liquid is described by the terms *low*, *medium*, and *high*, as shown in Fig. 1.5. The crisp input value $T = 50\,°C$ activates the two terms *low* and *medium* with the factors 0.8 and 0.2. The term *high* is not activated or is activated with the factor 0.

Note In some cases, it may be useful to convert the crisp input measured values into fuzzy values before they enter the fuzzy controller; the motivation for this lies in the fundamental tolerance inherent in every measurement and takes into account a subjective preliminary assessment of the tolerance intervals. This approach is useful if, for example, relatively inaccurate measuring transducers or analog-to-digital converters are selected for cost reasons. An example can be found in Sect. 8.1.

Approximate Reasoning

Linguistic variables are used to describe the system behavior by making a series of statements that define the corresponding output value combinations for certain input value combinations in the form of IF…THEN…rules. Let u and v be the two linguistic variables (e.g., "speed," "armature current " of an electric vehicle) and α and β be their terms (e.g., *large*, *quite small*). Then the linguistically formulated rules for system description can also be described in the following way by implications $(u = \alpha) \Rightarrow (v = \beta)$:

$$p : u \text{ is } \alpha$$

$$q : v \text{ is } \beta \tag{1.1}$$

$$\text{IF p THEN q} : (u = \alpha) \Rightarrow (v = \beta)$$

If specific (precise or imprecise) input values are stated for the controller, the individual statements must first be checked in order to calculate the output values. Since the input values will generally not exactly satisfy the conditions specified in the implications, only fuzzy conclusions or inferences can be drawn. The process required for this is called "approximate reasoning" and is derived from the methods of reasoning in binary logic. A simple method for implementing approximate reasoning is shown in Example 1.7. It is explained in more detail in Sect. 8.2.

Example 1.7 (Approximate Reasoning) The crisp control value φ_{OFF} for the opening of a cooling valve is to be adjusted based on measured crisp values T_{ON} of the input variable "temperature." Two rules are available in a linguistic form. The terms are represented by fuzzy sets as shown in Fig. 1.6. The rules for describing the controller behavior are as follows:

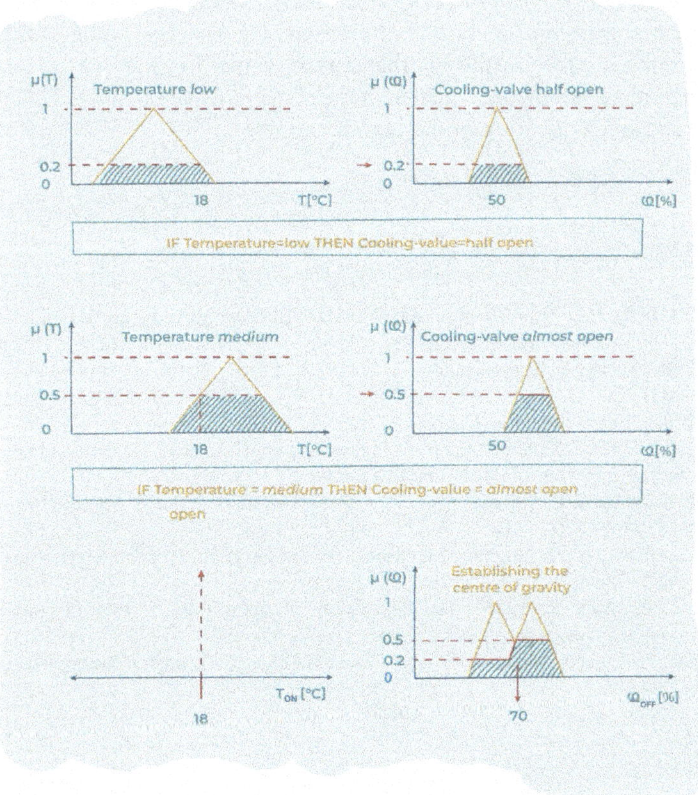

Fig. 1.6 Max-min inference method: The behavior of a fuzzy controller for setting the cooling valve opening φ_{OFF} as a function of a measured temperature T_{ON} is defined by two linguistically formulated rules. For an input value of 18 °C, an opening $\varphi = 70\%$ is set as the output value

IF temperature = low   THEN cooling valve = half open.
IF temperature = medium   THEN cooling valve = almost open.

The temperature $T_{\mathrm{ON}} = 18\,°\mathrm{C}$ is measured. To calculate the corresponding control value $\varphi_{\mathrm{OFF}}(18)$, the two membership values $\mu_{\mathrm{low}}(18\,°\mathrm{C}) = 0.2$ and $\mu_{\mathrm{medium}}(18\,°\mathrm{C}) = 0.5$ are first determined for the terms *low* and *medium* of the temperature. These are regarded as the resulting activation degrees of the rules and are used to fill in the two membership functions $\mu_{\mathrm{half\ open}}(\varphi)$ and $\mu_{\mathrm{almost\ open}}(\varphi)$ of the corresponding fuzzy terms of the cooling valve opening (if an activation degree is zero, the corresponding rule is said *not* to fire). The final fuzzy output value is obtained by superimposing the two individual results. It is represented by a fuzzy output set from which the required crisp control value value is obtained.

Defuzzification of the Output Variables
In order to generate crisp control values from the fuzzy initial sets for the individual control variables (as representatives of the fuzzy conclusions), the area under the curve of the membership function is often used. The abscissa value of the centroid of the area is used as the resulting crisp control value. In Example 1.7, this results in the crisp control value $\varphi_{\mathrm{OFF}} = 70\%$ of the cooling valve opening. This process is referred to as "defuzzification of the output variable."

References

Bocklisch, F. (1988). *Prozessanalyse mit unscharfen Verfahren*. VEB Verlag Technik. Berlin.

Bothe, H.-H. (1995). *Bewertung mit unscharfen Mengen*. Technische Hochschule Ilmenau.

Heite, R., Bothe, H.-H., & Zimmermann, H.-J. (1989). Fuzzy control: A survey. *Fuzzy Sets and Systems, 30*, 1–28.

Kosko, B. (1992). *Fuzzy thinking: The new science of fuzzy logic*. Hyperion.

Lukasiewicz, J. (1932). *Selected works* [Original work published in 1932]. North- Holland.

Preuß, R. (1992). *Fuzzy-control*. VEB Verlag Technik.

Surmann, H., Jorgensen, U. L., & Bunke, H. (1992). Fast fuzzy control for robotics. *Fuzzy Sets and Systems, 48*, 67–76.

Watanabe, T., Hirota, K., & Sugeno, M. (1990). A design of fuzzy controllers. *Fuzzy Sets and Systems, 30*, 1–28.

Yamakawa, T., & Miki, T. (1986). The application of fuzzy logic to control problems. *IEEE Transactions on Systems, Man, and Cybernetics, 16*, 45–52.

Zadeh. (1965). Fuzzy sets. *Information and Control, 8*, 338–353. https://doi.org/10.1016/S0019-9958(65)90241-X

Chapter 2
Crisp and Fuzzy Sets

Abstract This chapter focus on set theory, which was developed in the nineteenth century. Therein sets are defined as a collection of objects. They are introduced first and then extended to fuzzy sets. By allowing degrees of set memberships, they represent vagueness and imprecise information.

According to Cantor, a "set" combines certain objects or elements into a whole. Sets thus have an ordering or structuring character (e.g., see Bothe, 1995). In mathematics, the elements include numbers, number constructions (e.g., vectors, matrices, tensors, etc.), geometric structures, and mappings such as relations and functions. Particularly important sets are abbreviated with specific symbols (e.g., the sets of all natural, real, and complex numbers $\mathbb{N}$, $\mathbb{R}$, $\mathbb{C}$).

The distinction between a set and the elements not belonging to it can be either *crisp* or *fuzzy*. A crisply defined set M is understood as the collection of elements x that either belong to M or do not belong to M; the statement "x belongs to A" is either true or false. Conventional set theory is based on this concept of sets. An extension of this is fuzzy set theory (cf. Sects. 2.2–2.4).

2.1 Basic Concepts of Set Theory

Crisply defined sets are referred to below as "crisp sets" or simply "sets" and are written with symbols of normal thickness (e.g., A), while sets with vague boundaries are referred to as "fuzzy sets" and are written with symbols in bold ($\mathbf{A}$).

The elements of a set A can be specified by enumeration, by specifying a certain property, or graphically with the aid of diagrams.

Example 2.1 Set A with the five elements $a_1, a_2, \ldots, a_5$ can be represented by

$$A = \{a_1, a_2, \ldots, a_5\} = \{a_i \mid 1 \leq i \leq 5\}. \tag{2.1}$$

The set $A' = \{a_i \in A \mid a_i \geq 0.1\}$ selects all elements of A that are greater than 0.1.

H.-H. Bothe, E. Portmann, *Computing with Words*, Fuzzy Management Methods,
https://doi.org/10.1007/978-3-032-24117-7_2

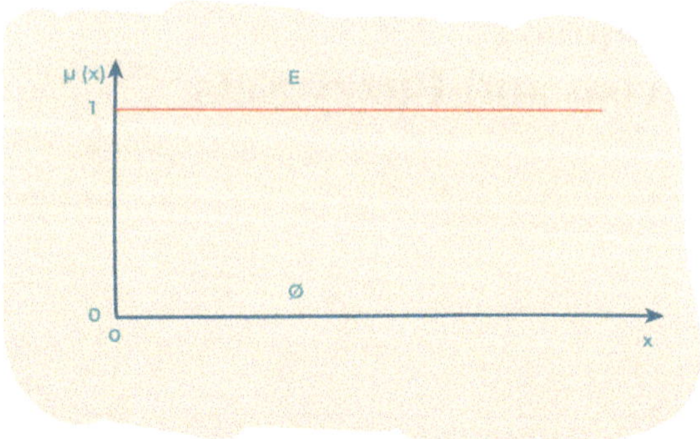

Fig. 2.1 Characteristic functions of the universal and empty sets, $\mu_E(x) = 1$ and $\mu_\varnothing(x) = 0$

If, in Example 2.1, the elements $a_1, \ldots, a_n$ are regarded as objects of a base set (or basic domain) X, then A can also be described by a 0–1-valued characteristic function $\mu_A : X \to \{0, 1\}$, for which $\forall x \in X$ applies:

$$
\mu_A(x) = \begin{cases} 1, & \text{if } x \in A, \\ 0, & \text{otherwise.} \end{cases} \tag{2.2}
$$

A function value $\mu_A(x_i) = 1$ marks the special elements $x_i \in A$ among all elements $x \in X$. In this case, A is a "subset" of X, written $A \subseteq X$. We also say that A is contained in X. The representations (2.1) and (2.2) are equivalent and can be easily converted into each other.

Definition 2.1 (Equality of Sets) Two sets A and B are called *equal*, written $A = B$, if they contain the same elements (i.e., if the curves of the two characteristic functions $\mu_A(x)$ and $\mu_B(x)$ are the same). If they are equal, $\forall x \in X$ applies:

$$
\mu_A(x) = \mu_B(x).
$$

Definition 2.2 (Universal Set, Empty Set) A set A is called the *universal set*, written $A = E$, if it is equal to the base set X, and the *empty set*, written $A = \varnothing$, if it contains no elements. The following applies: $\mu_E(x) = 1$ and $\mu_\varnothing(x) = 0$ (Fig. 2.1).

Sets can also be combined into higher-level sets or set systems. A special set system is the power set of a base set X.

Definition 2.3 (Power Set) The power set $P(X)$ on X is the set of all possible distinct subsets of X:

$$
P(X) := \{ A \mid A \subseteq X \}.
$$

The power set of a set with n elements thus contains 2^n possible sets as elements.

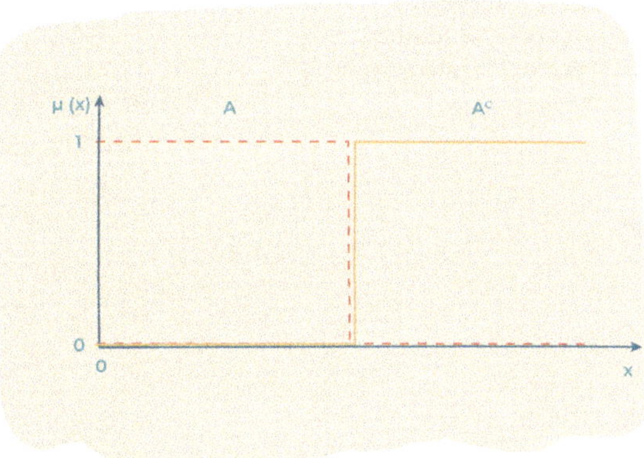

Fig. 2.2 Characteristic functions of a set A (dashed) and its complement A^C (bold)

Combinational operations are defined on sets, which are closely related to propositional logic and Boolean algebra. The basic operations are *complementation, intersection,* and *union.*

Definition 2.4 (Complement of a Set) Let X be a base set on which a set A is defined. Then the set $A^C = X \setminus A$ (read: X without A) is called the complement of A on X. For the characteristic function $\mu_{AC}(x)$, this means $\forall x \in \mathbf{X}$:

$$\mu_{AC}(x) = \begin{cases} 0 \\ 1 \end{cases} \quad \text{for} \quad \begin{array}{l} \mu_A(x) = 1, \\ \mu_A(x) = 0. \end{array}$$

This formula can be written in closed form as follows (Fig. 2.2):

$$\mu_{AC}(x) = 1 - \mu_A(x) \quad \forall x \in \mathbf{X}.$$

Definition 2.5 (Intersection of Two Sets) The intersection D of two sets A and B of the same domain X, written as $D = A \cap B$, consists of all elements $x \in X$ that belong to both sets at the same time. For $\forall x \in \mathbf{X}$, the following applies (Fig. 2.3):

$$\mu_{A \cap B}(x) = \begin{cases} 1 \\ 0 \end{cases} \quad \text{for} \quad \begin{array}{l} \mu_A(x) = \mu_B(x) = 1 \\ \text{otherwise.} \end{array}$$

There are several alternative ways to denote the characteristic function of the intersection in closed form, for example:

$$\mu_{A \cap B}(x) = \min\left[\mu_A(x), \mu_B(x) \right] = \mu_A(x) \cdot \mu_B(x) \quad \forall x \in \mathbf{X}.$$

Definition 2.6 (Union of Two Sets) The union V of two sets A and B of the same domain X, written $A \cup B$, consists of all elements $x \in X$ that belong to at least one of the two sets A or B. For $\forall x \in \mathbf{X}$, the following applies (Fig. 2.4):

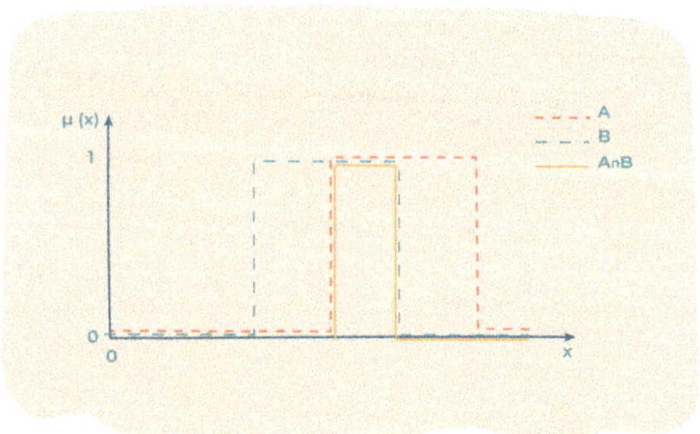

Fig. 2.3 Characteristic functions of A, B (dashed lines) and of the intersection $A \cap B$

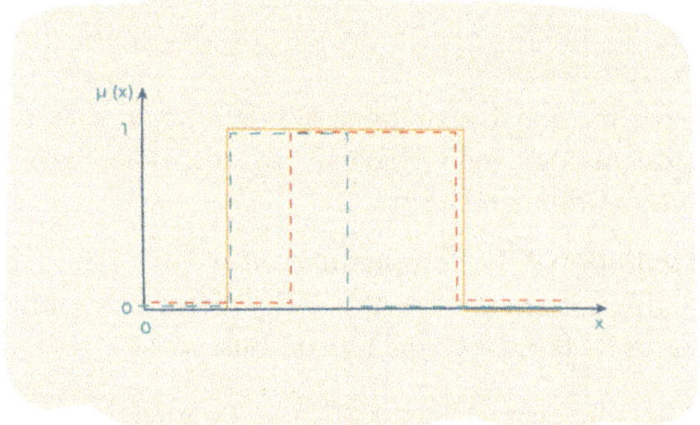

Fig. 2.4 Characteristic functions of A, B (dashed) and their union $A \cup B$

$$\mu_{A \cup B}(x) = \begin{cases} 1 \\ 0 \end{cases} \quad \text{for} \quad \begin{array}{l} \mu_A(x) = \mu_B(x) = 0 \\ \text{otherwise.} \end{array}$$

There are several ways to represent this equation in closed form, for example:

$$\mu_{A \cup B}(x) = \max\big[\mu_A(x),\, \mu_B(x)\big]$$
$$= \min\big[1,\, \mu_A(x) + \mu_B(x)\big]$$
$$= \mu_A(x) + \mu_B(x) - \mu_A(x) \cdot \mu_B(x) \quad \forall x \in \mathbf{X}$$

The alternative notations for forming the intersection and union of crisply defined sets are equivalent and yield the same result—in contrast to fuzzy sets (as we will see later). The diagrams according to Venn-Euler (see Fig. 2.5) provide a graphical illustration of complement formation, intersection, and union. The elements of the base set lie in a two-dimensional space, while the elements belonging to set A are held together by a closed curve.

Applications of conventional set theory in algebra arise, for example, when solving inequalities. The following example illustrates this.

Example 2.2 The solution sets A and B of the two inequalities $x > 4$ and $x < 14$ on the base set $\mathbb{R}$ are given by:

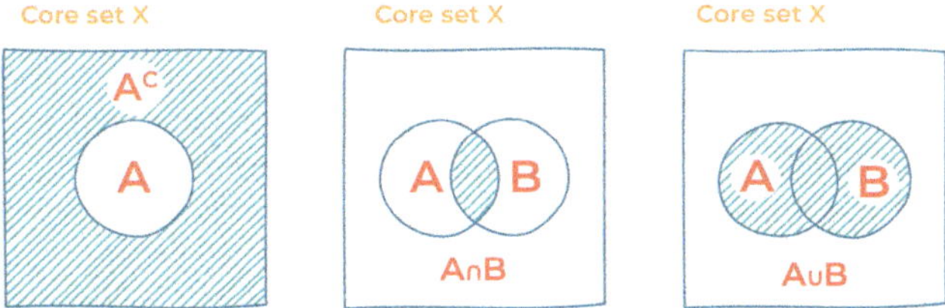

Fig. 2.5 Venn diagrams

$$A = \{x \in \mathbb{R} \mid x > 4\} \quad \text{and} \quad B = \{x \in \mathbb{R} \mid x < 14\}.$$

Based on the equivalence $(x > 4 \ \wedge \ x < 14) \iff (4 < x < 14)$, the solution set V of the intersection of A and B can be represented as:

$$V = \{x \in \mathbb{R} \mid 4 < x < 14\}.$$

In order to be able to compare sets that are defined on different bases, the concept of the "Cartesian product" is introduced. This forms the basis for a general definition of the concept of "relation." With the help of the Cartesian product, sets with different physical units can be structured. The Cartesian product maps a number n of different variables onto an n-dimensional space whose elements are n-tuples.

Definition 2.7 (Cartesian Product) The Cartesian product K of the sets $A_1, \ldots, A_n$, written as $K = A_1 \times \cdots \times A_n$, is defined by:

$$K = \{(x_1, \ldots, x_n) \mid x_i \in A_i\}.$$

All n-tuples possible by combining the elements of $A_1, \ldots, A_n$ are marked. This is also referred to as the *product space* of the n-tuples.

If the sets $A_1, \ldots, A_n$ are regarded as subsets of the base sets $X_1, \ldots, X_n$ with the characteristic functions $\mu_1(x_1), \ldots, \mu_n(x_n)$, then the characteristic function $\mu_K(x_1, \ldots, x_n)$ of K can be represented as:

$$\mu_K(x_1, \ldots, x_n) = \begin{cases} 1 \\ 0 \end{cases} \quad \text{for} \quad \begin{array}{l} \mu_1(x_1) = \ldots = \mu_n(x_n) = 1, \\ \text{otherwise.} \end{array}$$

The following formulas are suitable for closed representation:

$$\mu_K(x_1, \ldots, x_n) = \min\big[\mu_1(x_1), \ldots, \mu_n(x_n)\big] = \mu_1(x_1) \cdot (\ldots) \cdot \mu_n(x_n). \qquad (2.3)$$

In the case of a common base set X of $A_1, \ldots, A_n$, the numerical values $\mu_K(x_1, \ldots, x_n)$ are calculated in the same way as the numerical values $\mu_\cap(X_i)$ of the average intersection of $A_1, \ldots, A_n$.

For $n = 2$, the Cartesian product K represents a set of value pairs $(x_1, x_2) \in A_1 \times A_2$. If A_1 and A_2 consist of countably many elements x_1 and x_2, then $\mu_K(x_1, x_2)$ can also be represented in matrix notation. Only the value pairs with $\mu_K(x_1, x_2) = 1$ are taken into account in the product space of the base sets.

Example 2.3 Let the two sets $A_1, A_2 \subset \mathbb{N}$ be given with $A_1 = \{1, 5, 10\}$ and $A_2 = \{2, 6\}$. The Cartesian product $K = A_1 \times A_2$ is:

$$K = \{(1, 2), (5, 2), (10, 2), (1, 6), (5, 6), (10, 6)\}.$$

The characteristic function $\mu_K(x_1, x_2)$ can thus be represented as a matrix as follows:

$\mu_K(x_1, x_2)$: x_1 $\backslash$ x_2	2	6
1	1	1
5	1	1
10	1	1

The values of $x_1 \in \mathbb{N}\backslash A_1$ and $x_2 \in \mathbb{N}\backslash A_2$ (not shown in the matrix) are assigned the membership values $\mu_K(x_1', x_2') = 0$.

Definition 2.8 (Relation) The subsets $R \subseteq A_1 \times \cdots \times A_n$ of the Cartesian product of the sets $A_1, \ldots, A_n$ are called n-ary relations on $A_1 \times \cdots \times A_n$. The type of element selection is linked to a condition typical for relations. With $\mu_R(x_1, \ldots, x_n) \in \{0, 1\}$, the following always applies

$$\mu_R(x_1, \ldots, x_n) \leq \mu_K(x_1, \ldots, x_n).$$

Of particular importance are the two-place relations $R \subseteq A \times B$, which assign the elements $x \in A$ to the elements $y \in B$. Examples of this are the relations "$\leq$" and "$=$" between real numbers and "$\subseteq$" in set systems. If A and B contain countable elements, then the characteristic function of the mediating relation $R \subseteq A \times B$ can also be represented as a matrix. Relations are a generalized notation for equations and inequalities.

Example 2.4 Let the two sets $A_1 = \{1, 5, 10\}$ and $A_2 = \{2, 6\}$ be given with the Cartesian product according to Example 1.5. The relation $\mathbf{R} = \big\{ (x, y) \in A_1 \times A_2 \mid x \leq y \big\}$ is calculated as:

$$R = \{(1, 2), (1, 6), (5, 6)\}.$$

The characteristic function $\mu_R(X_1, X_2)$ can be represented in matrix form as:

$$\mu_R(x_1, x_2): \quad \begin{array}{c|cc} x_1 \backslash x_2 & 2 & 6 \\ \hline 1 & 1 & 1 \\ 5 & 0 & 1 \\ 10 & 0 & 0 \end{array}$$

If the second component of a two-place relation $R(x, y)$ is also the first component of another relation $S(y, z)$, then both relations can be concatenated or executed one after the other. The result is a relation $T(x, z) = R \circ S$ between the first component of $R(x, y)$ and the second component of $S(y, z)$. This referred to as a "composition" of relations.

Definition 2.9 (Concatenation of Relations) Let $R \subseteq A \times B$ and $S \subseteq B \times C$ be the two relations between the elements $x \in A$, $y \in B$, and $z \in C$. Then, concatenating S and R produces a relation $R \circ S \subseteq A \times C$ as the set of those value pairs (x, z) for which at least one value y exists with $(x, y) \in R \wedge (y, z) \in S$. When calculating the characteristic function $\mu_{R \circ S}(x, z)$, this means that for all value pairs (x, z) and all $y \in B$, it must be checked whether the values $\mu_R(x, y)$ and $\mu_S(y, z)$ exist. The characteristic function $\mu_{R \circ S}(x, z)$ is thus defined as:

$$\mu_{R \circ S}(x, z) = \sup_{y \in B} \left[\mu_R(x, y) \cdot \mu_S(y, z) \right].$$

The supremum $\sup_{y \in B}[\,]$ represents the smallest upper bound of the argument function for all $y \in B$ and can be replaced by the largest of these elements in the case of a countable number of elements. Since characteristic functions only take the values 0 or 1, multiplication can be replaced by a minimum formation in the defining equation according to Definition 2.9 without changing the result. This yields to:

$$\mu_{R \circ S}(x, z) = \sup_{y \in B} \left[\min \left[\mu_R(x, y), \ \mu_S(y, z) \right] \right]. \tag{2.4}$$

If sets A, B, and C contain countable elements and $\mu_R(x, y)$ and $\mu_S(y, z)$ can therefore be represented as matrices, then $\mu_{R \circ S}(x, z)$ can be calculated using the scheme for matrix multiplication, whereby addition must be replaced by the formation of the supremum.

Example 2.5 Let the sets be $A = \{1, 5, 10\}$, $B = \{0, 4, 8\}$, and $C = \{3.5, 4.5\}$ with elements $x \in A$, $y \in B$, and $z \in C$ as well as the relations *less than or equal to* $S = \{(y, z) \mid y \leq z\}$ and *less than* $R = \{(x, y) \mid x < y\}$ on $A \times B$ and $B \times C$, respectively. S, R, and $R \circ S$ then result in:

$$R(x, y) = \{(1, 4), (1, 8), (5, 8)\},$$

$$S(y, z) = \{(0, 3.5), (0, 4.5), (4, 4.5)\},$$

$$R \circ S(x, z) = \{(x, z) \mid \exists y \in B; \ x < y \wedge y \leq z\} = \{(1, 4.5)\}.$$

In matrix notation, this means:

$$
\begin{array}{c|cc}
\mu_S(y,z): & y^z & 3.5 & 4.5 \\
\hline
 & 0 & 1 & 1 \\
 & 4 & 0 & 1 \\
 & 8 & 0 & 0 \\
\end{array}
$$

$$
\begin{array}{c|ccc}
\mu_R(x,y): & x^y & 0 & 4 & 8 \\
\hline
 & 1 & 0 & 1 & 1 \\
 & 5 & 0 & 0 & 1 \\
 & 10 & 0 & 0 & 0 \\
\end{array}
\qquad
\begin{array}{c|cc}
\mu_{R\circ S}(x,z): & x^z & 3.5 & 4.5 \\
\hline
 & 1 & 0 & 1 \\
 & 5 & 0 & 0 \\
 & 10 & 0 & 0 \\
\end{array}
$$

2.2 Fuzzy Set Theory

The conventional working method of an engineer consists of finding the most accurate idealization of reality possibly using a fixed set of mathematical tools (e.g., in model building for system description). When fuzzy methods are used, this uniform mathematical apparatus is abandoned. Instead, heuristic elements are used to adapt the tools to the task at hand. This is essentially done in three steps:

1. Definition of problem-specific statements (based on expert knowledge)
2. Designing membership functions
3. Linking the individual statements and drawing conclusions (inference)

There are several different axiomatic systems in the field of fuzzy sets. Rules and laws vary depending on the current selection of the links used, whereby the requirement of consistency is often not fulfilled. The entirety of known axiomatic systems can be referred to as "fuzzy set theory."

The definition of fuzzy sets goes back to Zadeh (1965) and is uniform except for the notation. In the following, only the usual set notation is used, where the elements are written as pairs of values in round brackets and separated by a semicolon.

Definition 2.10 (Fuzzy Set) Let X be a set of elements or objects x that are to be evaluated for membership with regard to a fuzzy statement with a truth value $\mu_A(x)$. Then set A of value pairs $(x; \mu_A(x))$ with

$$A = \{(x, \mu_A(x)) \mid x \in X,\ \mu_A(x) \in R\}$$

is a fuzzy set on X with the membership function $\mu_A(x)$. The X is called the *basic domain* or *base set* of A.

Example 2.6 The set of real numbers approximately equal to 8 can be represented as:

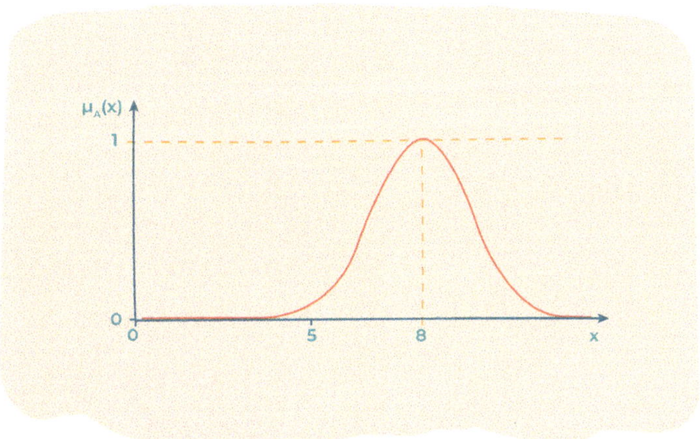

Fig. 2.6 Membership function $\mu_A(x) = [1 + (x - 8)^4]^{-1}$

$$A = \{(x, \mu_A(x)) \mid \mu_A(x) = [1 + (x - 8)^4]^{-1}\}.$$

The membership curve $\mu_A(x)$ is then shown in Fig. 2.6.

Example 2.7 For capacity reasons, a high-performance computer factory can produce at least four and at most nine computers of a certain type per working day. The possible daily output can therefore be clearly defined and represented by the set $A = \{4, 5, 6, 7, 8, 9\}$ of the daily production quantities. An expert could evaluate this set according to the criterion of *reasonable costs* as shown in Fig. 2.7:

$$A = \{(4; 0), \ (5; 0.1), \ (6; 0.5), \ (7; 1), \ (8; 0.8), \ (9; 0)\}.$$

Comments on Membership Functions $\mu(x)$

1. A membership function $\mu(x)$ should fulfil the following properties:

   - $\forall x \in X : \mu(x) \geq 0$
   - $\mu(x)$ is greater the better x fulfils the expert's evaluation criterion (in Example 2.7: reasonable costs).

2. The value range of the membership functions $\mu(x)$ that occur in completed problems should, but does not necessarily have to, be normalized (e.g., to prevent information loss). A number of different methods are available for this purpose, for example:

   - Unit interval normalization for $\mu(x)$: $x \rightarrow [0, 1]$ (2.2)
   - Normalization to a reference element x_o with $\mu_A(x_o) = 1$ (2.3)

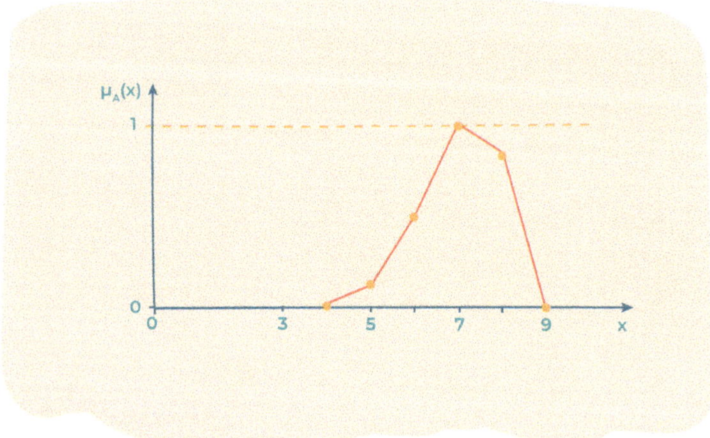

Fig. 2.7 Daily production at reasonable costs. The solid line between the discrete states of the abscissa serves to better visualize the function

In practice, both interval normalization and normalization to the element with the maximum membership value are used.

3. It should also be noted that a defined evaluation criterion (e.g., reasonable costs) is always used to determine the membership of an element in a set. The precision of the evaluation criterion determines the crispness of the set (Chap. 5).

Construction of a Membership Function

The procedure for determining the membership function of a fuzzy set **A** for a specific problem can be carried out in two different ways (cf., Bocklisch, 1988). In both cases, the elements $x \in X$ are evaluated by an expert in terms of a defined fuzzy criterion according to their membership:

1. Starting from a crisp subset $A' \subseteq X$, it is assumed that the membership of the boundary elements to set A is lower than that of the central elements and that the boundary elements have an influence beyond the (crisp) boundaries on neighboring elements lying outside. These thus also receive a certain membership. To model $\mu_A(x)$, a boundary level α can first be defined, and the membership curve $\mu_A(x)$ can be modeled using known and estimated system properties (Fig. 2.8).

2. All elements or objects $x \in X$ are known a priori. For example, all settings of a system are known so that they can be evaluated. Neighborhood relationships in the form of radiation models are not required. The elements can be discrete (Fig. 2.9).

In the following, some special fuzzy sets and important properties of fuzzy sets will be presented.

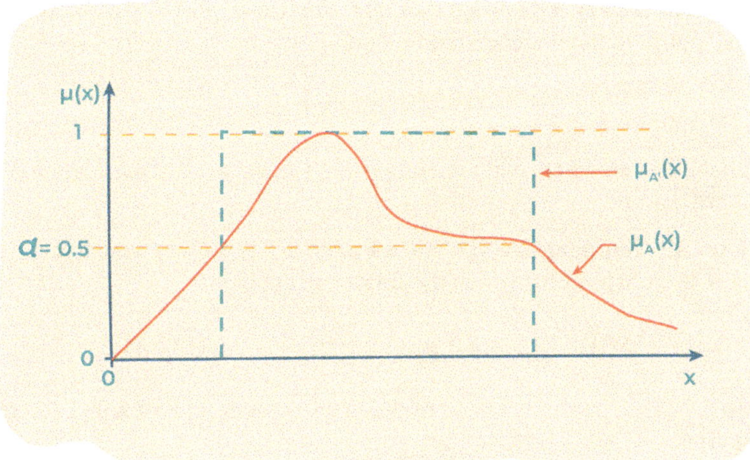

Fig. 2.8 Starting from an estimated crisp set A', a realistic membership function $\mu_A(x)$ can be constructed by allowing neighborhood relationships between the individual elements x to deform the crisp curve. The neighborhood relationships can result, for example, from stochastic system knowledge as probability distributions or from a subjective assessment

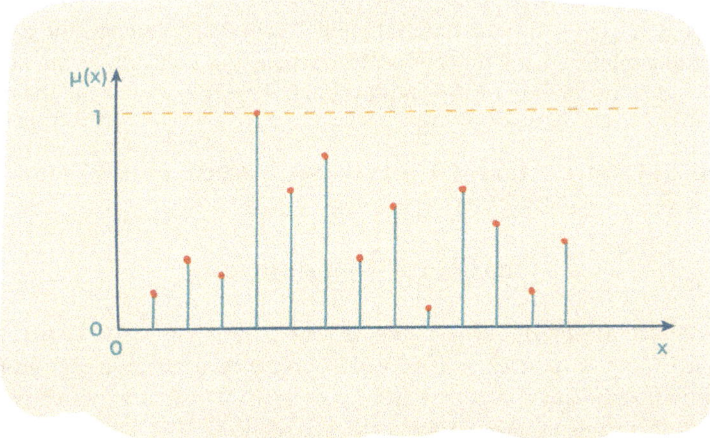

Fig. 2.9 Construction of a membership function: A large or representative selection of discrete elements is known and can be evaluated directly. If it is necessary to evaluate intermediate values, this can be done by interpolation

Definition 2.11 (Fuzzy Empty Set) The following applies to the membership function of the fuzzy empty set $\varnothing$:

$$\mu_\varnothing(x) = 0 \quad \forall x \in X.$$

Definition 2.12 (Fuzzy Universal Set for the Case $\mu_E(x) : X \to [0, 1]$) The
following applies to the membership function of the fuzzy universal set E:

$$\mu_E(x) = 1 \quad \forall x \in X.$$

To highlight the elements with membership values > 0, the definition of the
"support set" is used.

Definition 2.13 (Support Set S(A) of a Fuzzy Set A) The support set $S(A)$ of a
fuzzy set A is a crisp set for which the following applies[1]:

$$S(A) = \{x \in X \mid \mu_A(x) > 0\}, \qquad S(A) \subseteq X_1$$

Example 2.8 The support set $S(A)$ of the fuzzy set A from Example 2.7 (computer
production) is:

$$S(A) = \{5, 6, 7, 8\}.$$

Note If the support set $S(A)$ contains exactly one element $x_0 \in \mathbb{R}$ and this has the
membership value $\mu_A(x_0) = 1$, this is also referred to as the "singleton" x_0. A is
then a crisp set that corresponds to the numerical value x_0.

Sometimes it makes sense to highlight the "essential" elements of a fuzzy set
with a certain minimum membership. To do this, the concept of the support set
is generalized by introducing a threshold value α. The fuzzy set is mapped to a crisp
set via α.

Definition 2.14 (Intersection of A(α-Level Set, α-Cut)) Let A be a fuzzy set with
$A = \{(x; \mu_A(x)), x \in X\}$. Then

$$A_\alpha = \{x \in X \mid \mu_A(x) \geq \alpha\},$$

is the α-intersection of A. A_α is a crisp set with elements $x \in X$, for which the
membership function $\mu_A(x) \geq \alpha$ is valid, with α as a positive real number. For
normalization with $\mu_A(x) : X \to [0, 1]$, $\alpha \in [0, 1]$ must apply. If $\forall x \in X :$
$\mu_A(x) > \alpha$, this is also referred to as a strict α-section.

Example 2.9 The fuzzy set $A = \{(x; \mu_A(x)), x \in X\}$ holds as shown in Fig. 2.10.
The intersections of A are then plotted according to the marked intervals of the
numerical value scale below.

For strict intersections, the relationship from Theorem 2.1 applies. Example 2.9
serves to illustrate this.

---

[1] The support set of the fuzzy empty set is the empty set.

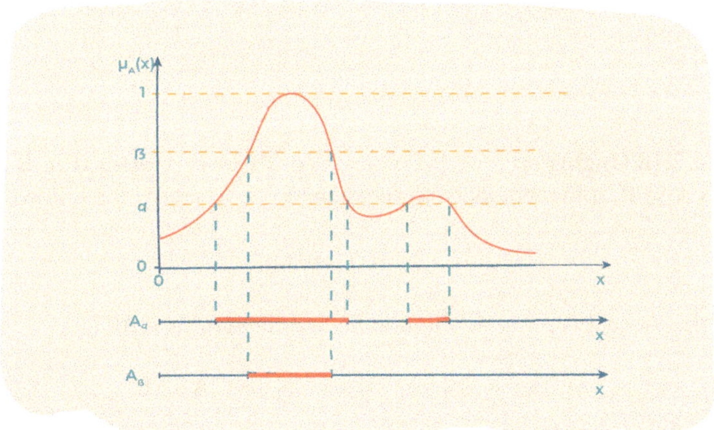

Fig. 2.10 Intersections of a fuzzy set **A** at membership levels α and β

Theorem 2.1

$$\alpha < \beta \Rightarrow A_\beta \subset A_\alpha \tag{2.5}$$

According to Sect. 2.1, crisp sets can in turn be combined into (super) sets whose elements are precisely these sets. Transferring this to fuzzy sets leads to the concept of "fuzzy power sets."

Definition 2.15 (Fuzzy Power Set) The set of all fuzzy sets **A** on a crisp base set X is called the fuzzy power set of X, written **P(X)**.

Further transfers from conventional set theory follow, which make statements about special properties of the fuzzy set under consideration.

Definition 2.16 (Height) The smallest upper bound of $\mu_A(x)$ on X, (i.e., the supremum of $\mu_A(x)$), is called the height hgt $(\mathbf{A})$. The following applies:

$$\text{hgt}(\mathbf{A}) = \sup_{x \in X}\left[\,\mu_A(x)\,\right].$$

The height of **A** denotes the element x_0 from **A** with the largest membership value $\mu_A(x_0)$. When normalized to this element, hgt $(\mathbf{A}) = 1$. Normalization can therefore be achieved by dividing the membership function $\mu_A(x)$ by $\mu_A(x_0)$.

Example 2.10 Given $\mathbf{A} = \{(1; 0.2), (2; 0.5), (4; 0.8), (6; 0.3)\}$. With hgt$(A) = 0.8$, the following applies:

$$\mathbf{A}_{\text{norm}} = \{(1; 0.25), (2; 0.625), (4; 1), (6; 0.375)\}.$$

Definition 2.17 (Equality) Two fuzzy sets $\mathbf{A}, \mathbf{B} \in \mathbf{P}(X)$ are equal, written $\mathbf{A} = \mathbf{B}$, if $\forall x \in X$ holds:

$$\mu_A(x) = \mu_B(x).$$

Definition 2.18 (Inclusion[2]) A fuzzy set $\mathbf{A} \in \mathbf{P}(X)$ is contained in $\mathbf{B} \in \mathbf{P}(X)$, written as $\mathbf{A} \subseteq \mathbf{B}$, if the membership functions $\forall x \in X$ satisfy the following:

$$\mu_A(x) \leq \mu_B(x).$$

With a strict inequality sign, A is said to be *truly contained* in $\mathbf{B}$:

$$\mu_A(x) < \mu_B(x), \quad \forall x \in X \quad \Leftrightarrow \quad \mathbf{A} \subset \mathbf{B}.$$

Example 2.11 Let $X = \{10, 20, 50, 100, 200\}$ be the set of possible measurement range settings $x \in X$ for a computer-controlled measuring device to be used in quality control. Three people, A, B, and C, make the following assessments in the form of fuzzy sets with respect to the fuzzy statement *x is a reasonable initial value setting* for a specific diagnostic situation:

$$\mathbf{A} = \{(10; 0), \ (20; 0.2), \ (50; 0.4), \ (100; 0.6)\},$$
$$\mathbf{B} = \{(10; 0), \ (20; 0.1), \ (50; 0.1), \ (100; 0.5)\},$$
$$\mathbf{C} = \{(10; 0.1), \ (20; 0.3), \ (50; 0.9), \ (100; 0.7)\}.$$

The following then applies:

$$\mathbf{B} \subseteq \mathbf{A} \subset \mathbf{C}.$$

According to Zimmermann (1991) and Rommelfanger (1988), the following theorem applies:

Theorem 2.2

$$1. \ \mathbf{A} \subseteq \mathbf{B} \wedge \mathbf{B} \subseteq \mathbf{A} \ \Leftrightarrow \ \mathbf{A} = \mathbf{B} \ [\text{Identity}] \tag{2.6}$$

$$2. \ \mathbf{A} \subseteq \mathbf{B} \wedge \mathbf{B} \subseteq \mathbf{C} \ \Leftrightarrow \ \mathbf{A} \subseteq \mathbf{B} \ [\text{Transitivity}] \tag{2.7}$$

$$3. \ \mathbf{A} \subseteq \mathbf{B} \Rightarrow \mathbf{S(A)} \subseteq \mathbf{S(B)} \tag{2.8}$$

$$4. \ \mathbf{A} \subseteq \mathbf{B} \Rightarrow \mathbf{A}_\alpha \subseteq \mathbf{B}_\alpha \tag{2.9}$$

$$5. \ \forall \mathbf{A} \in \mathbf{P}(X) : \ \varnothing \subseteq \mathbf{A} \tag{2.10}$$

---

[2] Compare classical set theory with characteristic function $\mu(x)$.

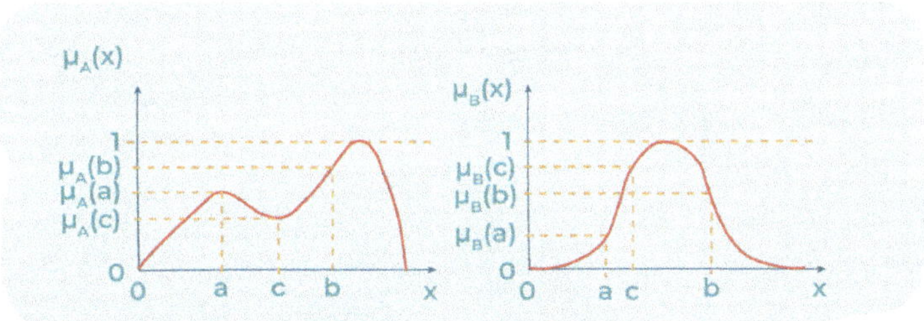

Fig. 2.11 Example of a non-convex (left) and a convex (right) fuzzy set

Definition 2.19 (Convexity) A fuzzy set $A = \{(x; \mu_A(x)), x \in X\}$ is called convex if the following applies to the entire domain of interest X:

$$\forall a, b, c \in X : \mu_A(c) \geq \min[\mu_A(a), \mu_A(b)] \quad \text{with } a \leq c \leq b.$$

Figure 2.11 shows membership functions for a non-convex fuzzy set **A** and a convex fuzzy set **B**

Comments
1. The term "convex fuzzy set" does not imply that $\mu_A(x)$ is a "convex function" in the conventional sense (left or right curving).
2. A fuzzy set **A** is convex if and only if all its intersections are convex in the set-theoretical sense (i.e., they are connected).

Definition 2.20 (Cardinality) Let X be a finite set. Then the cardinality of a fuzzy set $A \in \mathbf{P}(X)$ is defined as:

$$|A| = \sum_{x \in X} \mu_A(x).$$

The size $\|A\| = |A|/|X|$ is called relative cardinality.

Note Since X is a crisp set, the cardinality of the set of elements of $|X|$—with characteristic function not equal to 0—is equal to the number of elements in X. The relative cardinality of a fuzzy set therefore depends on the cardinality of the corresponding base set. In order to compare fuzzy sets using their relative cardinality, the same base set must therefore be assumed.

Example 2.12 Let $A = \{(5; 0.1), (6; 0.5), (7; 1), (8; 0.8)\}$. The following applies:

$$|A| = 2.4, \quad |X| = 4, \quad \|A\| = 2.4/4 = 0.6.$$

An extension of the concept of fuzzy sets was introduced by Zadeh (1973). He identified the fundamental difficulty of assigning a "crisp" membership function to

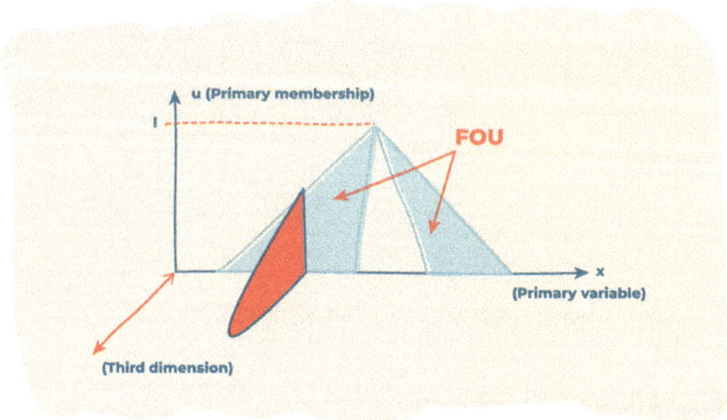

Fig. 2.12 Example membership function of a general type 2 fuzzy set in three (2+1) dimensions with a footprint of uncertainty (FOU), which can be thought of as fuzzifying a type 1 membership function (i.e., adding the third dimension)

a given statement. With the introduction of fuzzy membership functions, the fuzzy set of the second order is defined. Fuzzy sets according to Definition 2.10 are then referred to as fuzzy sets of the first order.

Definition 2.21 (Second-Order Fuzzy Set) A second-order fuzzy set is a fuzzy set whose membership values themselves form fuzzy sets in the interval [0, 1] (see Fig. 2.12).

As a generalization of Definition 2.21, fuzzy sets of order m can also be defined.

2.3 Operations on Fuzzy Sets

The following chapter presents a selection of the most important operations with fuzzy sets presented in the literature.

A very simple operation is complementation. For example, the statement "the steam locomotive is traveling slowly" is transformed into the statement "it is false that the steam locomotive is traveling slowly" (i.e., not necessarily into "the steam locomotive is traveling fast"; Example 1.5).

Definition 2.22 (Complement) Let $\mathbf{A} \in \mathbf{P}(X)$. If $\mu_A : X \rightarrow [0, 1]$, then $\mathbf{A}^C$ is called the complement of $\mathbf{A}$ on X with (Fig. 2.13):

$$\forall x \in X : \mu_{AC}(x) = 1 - \mu_A(x).$$

As can be easily verified, the following laws apply:

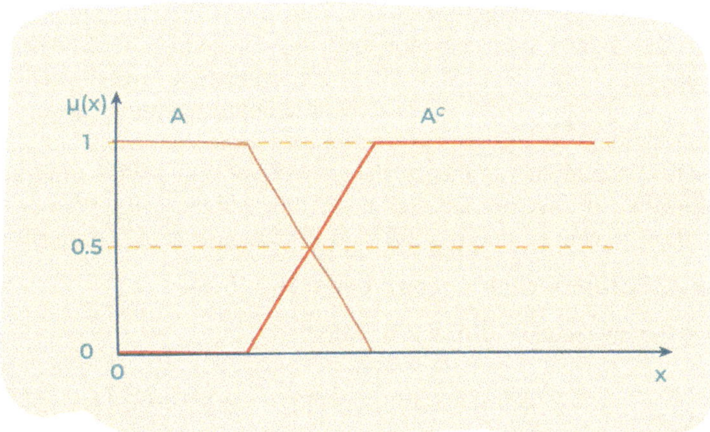

Fig. 2.13 Membership functions of a fuzzy set **A** and its complement **A**^{C}

$$[\mathbf{A}^C]^C = \mathbf{A} \tag{2.11}$$

$$\mathbf{A} \subseteq \mathbf{B} \;\Leftrightarrow\; \mathbf{B}^C \subseteq \mathbf{A}^C \tag{2.12}$$

Remark An extended definition of the complement leads to:

$$\forall x \in X : \; \mu_{AC}(x) = [1 - \mu_A(x)]^p \quad \text{with } p > 0.$$

In the following, some multi-digit operations will be described. A crisp base set X is evaluated according to different criteria A and B. These evaluations are represented by the fuzzy sets **A** and **B**. An overall evaluation will take both criteria into account and lead to a fuzzy result set **C**.

Example 2.13 (Computer Production According to Example 2.7) The evaluation of the number of possible daily units *at reasonable production costs* is given by the fuzzy set A as:

$$\mathbf{A} = \{(4; 0), (5; 0.1), (6; 0.5), (7; 1), (8; 0.8), (9; 0)\}.$$

The second assessment is carried out by the sales department based on the criterion of *daily sellability* and is given by the fuzzy set B as:

$$\mathbf{B} = \{(4; 1), (5; 0.9), (6; 0.8), (7; 0.4), (8; 0.1), (9; 0)\}.$$

This raises the question of how the cost side of daily production can be evaluated taking both criteria into account.

The most important linking operators for the verbal description of fuzzy information are the logical operators not, and, or. In analogy to Boolean algebra, conventional set theory is used to link the evaluations represented by fuzzy sets. While the operator "not" is realized by complementation, there are several possibilities for realizing the two basic operations, intersection and union, whereby, in contrast to conventional set theory, the alternatives now lead to different results. In the following, intersection and union are first defined using the min and max operators; the alternatives are given their own function-specific designations.

Definition 2.23 (Intersection, Union) Let $A, B \in P(X)$, then:

– $A \cap B$ is the intersection of A and B with:

$$\forall x \in X : \mu_{A \cap B}(x) = \min[\,\mu_A(x), \mu_B(x)\,],$$

– $A \cup B$ is the union of A and B with:

$$\forall x \in X : \mu_{A \cup B}(x) = \max[\,\mu_A(x), \mu_B(x)\,].$$

The intersection and union are formed point by point over the basic domain X.

Example 2.14 The complement, intersection, and union of two fuzzy sets A and B can be represented in Fig. 2.14 by the boundary curve of the hatched areas:

The following laws apply to the min and max operators:

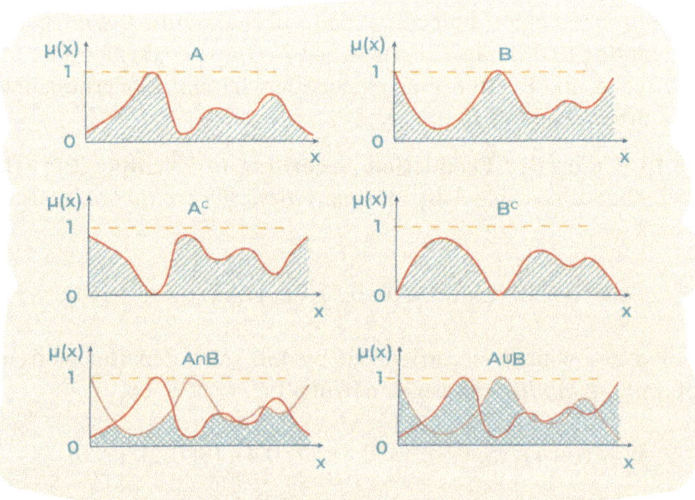

Fig. 2.14 Extended Venn diagrams for complement, intersection, and union (according to Zimmermann, 1991)

$$\text{Commutativity}: \quad \mathbf{A} \cap \mathbf{B} = \mathbf{B} \cap \mathbf{A},$$

$$\mathbf{A} \cup \mathbf{B} = \mathbf{B} \cup \mathbf{A}, \tag{2.13}$$

$$\text{Associativity:} \quad (\mathbf{A} \cap \mathbf{B}) \cap \mathbf{C} = \mathbf{A} \cap (\mathbf{B} \cap \mathbf{C}),$$

$$(\mathbf{A} \cup \mathbf{B}) \cup \mathbf{C} = \mathbf{A} \cup (\mathbf{B} \cup \mathbf{C}), \tag{2.14}$$

$$\text{Distributivity:} \quad \mathbf{A} \cap (\mathbf{B} \cup \mathbf{C}) = (\mathbf{A} \cap \mathbf{B}) \cup (\mathbf{A} \cap \mathbf{C}),$$

$$\mathbf{A} \cup (\mathbf{B} \cap \mathbf{C}) = (\mathbf{A} \cup \mathbf{B}) \cap (\mathbf{A} \cup \mathbf{C}), \tag{2.15}$$

$$\text{Adjunctivity:} \quad \mathbf{A} \cap (\mathbf{B} \cup \mathbf{A}) = \mathbf{A},$$

$$\mathbf{A} \cup (\mathbf{B} \cap \mathbf{A}) = \mathbf{A}, \tag{2.16}$$

$$\text{De Morgan's laws:} \quad [\mathbf{A} \cap \mathbf{B}]^{C} = \mathbf{A}^{C} \cup \mathbf{B}^{C},$$

$$[\mathbf{A} \cup \mathbf{B}]^{C} = \mathbf{A}^{C} \cap \mathbf{B}^{C}. \tag{2.17}$$

As can be easily verified, the law of complementarity does not apply to intersections and unions:

$$\mathbf{A} \cap \mathbf{A}^{C} \neq \varnothing \quad \text{and} \quad \mathbf{A} \cup \mathbf{A}^{C} \neq \mathbf{E} \tag{2.18}$$

Example 2.15 (Proof of the First De Morgan's Law) The following applies $\forall x \in X$:

$$\mu_{(A \cap B)^{C}}(x) = 1 - \min[\mu_A(x), \mu_B(x)],$$

$$\mu_{A^{C} \cup B^{C}}(x) = \max[1 - \mu_A(x), 1 - \mu_B(x)].$$

The second formula can be reformulated as:

$$1 - \mu_A(x) \geq 1 - \mu_B(x) \ \Leftrightarrow \ \mu_A(x) \leq \mu_B(x)$$

1. Case: $\Rightarrow \mu_{A^{C} \cup B^{C}}(x) = 1 - \mu_A(x),$

$$1 - \mu_A(x) < 1 - \mu_B(x) \ \Leftrightarrow \ \mu_B(x) < \mu_A(x)$$

2. Case: $\Rightarrow \mu_{A^{C} \cup B^{C}}(x) = 1 - \mu_B(x).$ So:

$$\mu_{A^{C} \cup B^{C}}(x) = 1 - \min[\mu_A(x), \mu_B(x)] = \mu_{(A \cap B)^{C}}(x). \qquad \text{q.e.d.}$$

Due to the good consistency between the axioms for the min/max operators and the laws of Boolean algebra, the terms "intersection" and "union" are justified in retrospect. When combining fuzzy sets, they embody the logical and ($\wedge$) or the logical or ($\vee$) of the corresponding fuzzy statements. A complete axiomatic system was established by Bellman and Giertz (1978).

Note Both operations obviously prohibit membership values $\mu_Z(x)$ between the minimum value and the maximum value at position x: Extreme evaluations cannot be balanced by moderate ones. Intersection and union only take into account one of the membership curves to be linked in each area. The variation of one of the fuzzy sets to be linked can therefore lead to the same linking result.

In addition to the min and max operators, alternative operators for intersection and union have been defined that are associated with other properties. With the help of a (reformulated) De Morgan's law (2.17), for example, a corresponding $\cap$-operator can be derived from a given $\cup$-operator. Let $\square$ be one of these intersection operators; then by

$$\mathbf{A} \blacksquare \mathbf{B} = \left[\mathbf{A}^C \square \mathbf{B}^C\right]^C , \tag{2.19}$$

it is assigned a corresponding union operator $\blacksquare$. To establish a systematic structure, both operators must satisfy additional minimum conditions, which are defined in the terms t-norm and s-norm . For this purpose, the operations are written as follows:

$$\mathbf{A} \square \mathbf{B} \;\rightarrow\; t(\mu_A(x), \mu_B(x)) , \tag{2.20}$$

$$\mathbf{A} \blacksquare \mathbf{B} \;\rightarrow\; s(\mu_A(x), \mu_B(x)) . \tag{2.21}$$

Definition 2.24 (T(riangular) Norm and S-Norm (T-Conorm))   A two-step function $t : [0, 1] \times [0, 1] \rightarrow [0, 1]$ is called a t-norm , if for $\forall x \in X$ the following conditions apply (abbreviated notation: $\mu = \mu(x)$):

1. $t(0, 0) = 0$ and $t(\mu_A, 1) = t(1, \mu_A) = \mu_A$.
2. $t(\mu_A, \mu_B) \leq t(\mu_C, \mu_D) \;\forall \mu_A \leq \mu_C, \; \mu_B \leq \mu_D$.
3. Commutativity : $t(\mu_A, \mu_B) = t(\mu_B, \mu_A)$.
4. Associativity: $t(\mu_A, t(\mu_B, \mu_C)) = t(t(\mu_A, \mu_B), \mu_C)$.

A two-step function $s : [0, 1] \times [0, 1] \rightarrow [0, 1]$ is called an s-norm , if for $\forall x \in X$ the following conditions apply (abbreviated notation: $\mu = \mu(x)$):

1. $s(1, 1) = 1$ and $s(\mu_A, 0) = s(0, \mu_A) = \mu_A$.
2. $s(\mu_A, \mu_B) \leq s(\mu_C, \mu_D) \;\forall \mu_A \leq \mu_C, \; \mu_B \leq \mu_D$.
3. Commutativity : $s(\mu_A, \mu_B) = s(\mu_B, \mu_A)$.
4. Associativity: $s(\mu_A, s(\mu_B, \mu_C)) = s(s(\mu_A, \mu_B), \mu_C)$.

The complementarity conditions $\left(t[\mu_A, 1 - \mu_A] = 0, \; s[\mu_A, 1 - \mu_A] = 0\right)$ do not necessarily have to be satisfied; however, a t-norm and the corresponding s-norm can be converted into each other on the basis of a rewritten De Morgan's law (2.19):

$$s[\mu_A(x), \mu_B(x)] = 1 - t(1 - \mu_A(x), 1 - \mu_B(x)) . \tag{2.22}$$

Compliance with the associativity law is particularly important, as this also enables recursive links. The min operator is an example of a t-norm , while the max operator is an example of an s-norm.

We will now introduce some combinations of operations. In contrast to the min and max operators, all numerical membership values are used here to calculate the resulting membership function.

Definition 2.25 (Algebraic Product) Let $A, B \in P(X)$. The algebraic product ($\hookrightarrow$ intersection) of A and B, written $A \cdot B$, is defined as a fuzzy set with:

$$\mu_{A \cdot B}(x) = \mu_A(x) \cdot \mu_B(x).$$

The membership function of the algebraic sum ($\hookrightarrow$ union) is then given by (2.19)

$$\mu_{A+B}(x) = 1 - \big(1 - \mu_A(x)\big).\big(1 - \mu_B(x)\big)$$
$$= \mu_A(x) + \mu_B(x) - \mu_A(x) \cdot \mu_B(x).$$

Due to the similarity of the formula to an addition theorem in probability theory (Sect. 5.1), this is also referred to as the probabilistic sum.

In addition to (2.19), the second De Morgan's law also applies to the algebraic operators:

$$\mu_{A^C \cdot B^C}(x) = \big(1 - \mu_A(x)\big).\big(1 - \mu_B(x)\big)$$
$$= 1 - \Big[\mu_A(x) + \mu_B(x) - \mu_A(x).\mu_B(x)\Big]$$
$$= \mu_{(A+B)^C}(x). \quad \text{q.e.d.}$$

However, like intersection and union, the algebraic operators ($\cdot$) and ($+$) are not complementary, since generally $\exists x \in X$ such that $\mu_A(x) \cdot (1 - \mu_A(x)) \neq 0$ applies.

Example 2.16 (Computer Production According to Example 2.13) Given the fuzzy sets (Fig. 2.15):

$$A = \{(4; 0),\ (5; 0.1),\ (6; 0.5),\ (7; 1),\ (8; 0.8),\ (9; 0)\},$$
$$B = \{(4; 1),\ (5; 0.9),\ (6; 0.8),\ (7; 0.4),\ (8; 0.1),\ (9; 0)\}.$$

Then:

$$A \cdot B = \{(4; 0),\ (5; 0.09),\ (6; 0.4),\ (7; 0.4),\ (8; 0.08),\ (9; 0)\},$$
$$A + B = \{(4; 1),\ (5; 0.91),\ (6; 0.9),\ (7; 1),\ (8; 0.82),\ (9; 0)\}.$$

Definition 2.26 (Bounded Product) Let $A, B \in P(X)$. The bounded product of A and B ($\hookrightarrow$ intersection), written $A \sqcap B$, is defined as a fuzzy set with:

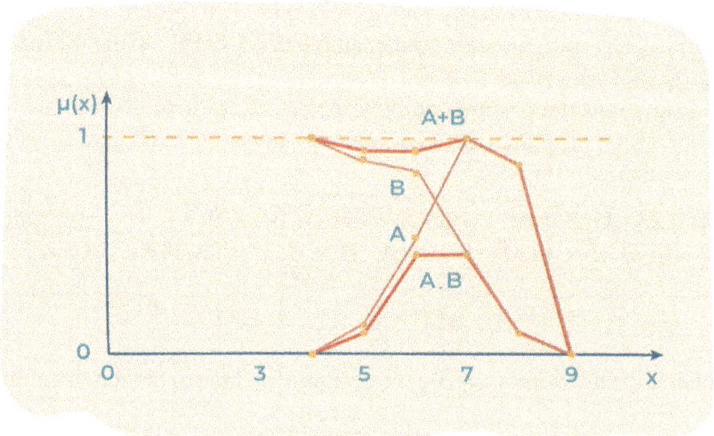

Fig. 2.15 Membership functions $\mu_{A\cdot B}(x)$ and $\mu_{A+B}(x)$

$$\forall x \in X: \ \mu_{A\sqcap B}(x) = \max\left[0, \ \mu_A(x) + \mu_B(x) - 1\right].$$

The membership function of the bounded sum of **A** and **B** ($\hookrightarrow$ union) is then given by (2.19) as:

$$\mu_{A\sqcup B}(x) = 1 - \max\left[0, \ (1 - \mu_A(x)) + (1 - \mu_B(x)) - 1\right]$$
$$= 1 - \max\left[0, \ 1 - \mu_A(x) - \mu_B(x)\right]. \tag{2.23}$$

This formula can be simplified by distinguishing between cases:

1. Case: $\quad \mu_A(x) + \mu_B(x) < 1 \ \Rightarrow \ 1 - (\mu_A(x) + \mu_B(x)) > 0$
$$\Rightarrow \ \mu_{A\sqcup B}(x) = \mu_A(x) + \mu_B(x).$$

2. Case: $\quad \mu_A(x) + \mu_B(x) > 1 \ \Rightarrow \ \mu_{A\sqcup B}(x) = 1.$

In both cases, the result is the smaller of the two values, $\mu_A(x) + \mu_B(x)$ and 1. This gives:

$$\mu_{A\sqcup B}(x) = \min\left[1, \ \mu_A(x) + \mu_B(x)\right]. \tag{2.24}$$

While the intersection is formed by a "true" multiplication of the membership values when defining the algebraic operators, according to (2.24) the union is achieved by adding the membership values for the bounded operators. To maintain normalization to [0, 1], it is necessary to restrict the membership values using $\min[1, \cdot]$.

The bounded operators are commutative and associative (see 2.13 and 2.14), but not distributive or adjunctive (2.15, 2.16). In contrast to intersection and union, they satisfy the law of complementarity. The following theorem applies:

Theorem 2.3 $\ \mathbf{A} \sqcap \mathbf{A}^{\mathbf{C}} = \varnothing \ \wedge \ \mathbf{A} \sqcup \mathbf{A}^{\mathbf{C}} = \mathbf{E}.$

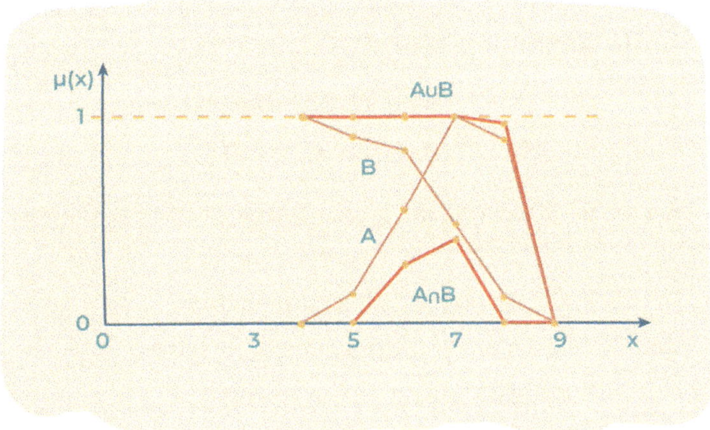

Fig. 2.16 Membership functions $\mu_{A\sqcap B}(x)$ and $\mu_{A\sqcup B}(x)$

Proof $\forall x \in X: \ \mu_{A\sqcap A^c}(x) = \max[\,0, \ \mu_A(x) + (1 - \mu_A(x)) - 1\,] = 0,$
$$\mu_{A\sqcup A^c}(x) = \min[\,1, \ \mu_A(x) + (1 - \mu_A(x))\,] = 1. \quad \text{q.e.d.}$$

Example 2.17 With the fuzzy sets **A** and **B** from Example 2.13 (computer production), the following union results are obtained:

$$A \sqcap B = \{(4; 0), \ (5; 0), \ (6; 0.3), \ (7; 0.4), \ (8; 0), \ (9; 0)\},$$

$$A \sqcup B = \{(4; 1), \ (5; 1), \ (6; 1), \ (7; 1), \ (8; 0.9), \ (9; 0)\}.$$

The combination results are plotted in the following diagram (Fig. 2.16):

A comparison of the operators discussed so far leads to the following theorem:

Theorem 2.4

$$\mathbf{A \sqcap B \subseteq A \cdot B \subseteq A \sqcap B} \quad \text{and} \quad \mathbf{A \sqcup B \supseteq A + B \supseteq A \sqcup B}$$

The curves of the resulting membership functions are therefore always below ($\bigcap$) or above ($\bigcup$) the following in the series. When formalizing decision-making using fuzzy sets, the result may depend on the type of operators used. This is illustrated by the following example.

Example 2.18 Evaluation of the performance of computers according to the criteria:

    A: Computing power for floating point operations in small loops

    B: Fast hard disk access

There are three devices to choose from, which are numbered for anonymization: $X = \{1, 2, 3\}$. The assessment is as follows:

$$A = \{(1; 0.9), (2; 0.5), (3; 0.1)\},$$

$$B = \{(1; 0.3), (2; 0.5), (3; 0.9)\}.$$

If both criteria are to be met ($\wedge$), the overall assessments are as follows, based on the intersection values discussed:

$$A \cap B = \{(1; 0.3), (2; 0.5), (3; 0.1)\},$$

$$A \cdot B = \{(1; 0.27), (2; 0.25), (3; 0.09)\},$$

$$A \sqcap B = \{(1; 0.2), (2; 0), (3; 0)\}.$$

If at least one criterion must be optimally fulfilled ($\vee$), the overall assessments are obtained by the union of the two partial statements:

$$A \cup B = \{(1; 0.9), (2; 0.5), (3; 0.9)\},$$

$$A + B = \{(1; 0.93), (2; 0.75), (3; 0.91)\},$$

$$A \sqcup B = \{(1; 1), (2; 1), (3; 1)\}.$$

The different operators therefore lead to different purchase recommendations. When deciding on a specific result, the properties of the operators must be taken into account; for example, ($\cup$) and ($\cap$) only consider the most positive or most negative partial result. The intersection operators tend to lead to a compromise between the two specifications, while the union operators, with the exception of ($\sqcup$), tend to evaluate the presence of at least one positive property. It must also be examined whether both criteria are actually equally important. Defining a specific operation for linking or "aggregating" partial results to form the overall result is an essential task in decision-making. This is described in more detail in Chaps. 7, 8, and 9.

The intersection operators described so far necessarily lead to a correspondingly negative result in the overall evaluation if an element of the support set is evaluated too negatively with regard to a partial aspect. No correction or compensation by good evaluations based on other partial aspects is possible. The operators are too pessimistic in their assessment. Accordingly, the union operators described so far are too optimistic in their assessment.

When making an overall assessment based on competing (or contradictory) statements, a compromise must obviously be reached. Since the partial assessments of the individual aspects are based on the subjective view of an expert, in many cases it is advisable to move away from a strict logical "and" (or "or") in favor of operators that lie between the intersection and the union. To this end, let us first define the arithmetic and geometric means of two fuzzy sets.

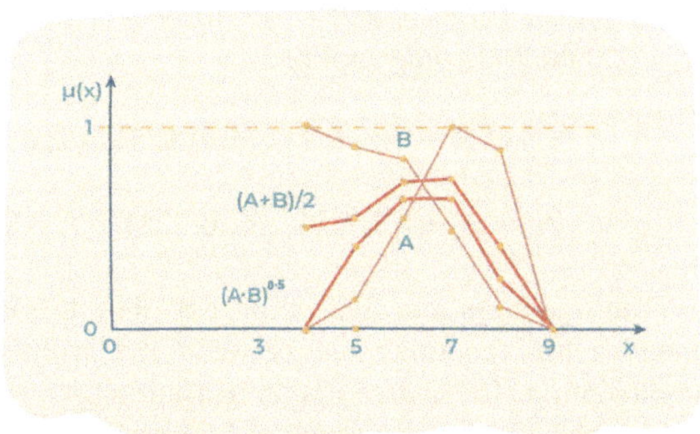

Fig. 2.17 Membership functions for the arithmetic and geometric mean values with specified values according to Example 2.13

Definition 2.27 (Arithmetic Mean) Let $\mathbf{A}, \mathbf{B} \in P(X)$ be given. Then the fuzzy set $\frac{1}{2}(A + B)$ is the arithmetic mean of $\mathbf{A}$ and $\mathbf{B}$ on X, where the following applies for all $x \in X$:

$$\mu_{\frac{1}{2}(A+B)}(x) = \frac{1}{2}\left[\mu_A(x) + \mu_B(x)\right].$$

Definition 2.28 (Geometric Mean) Let $\mathbf{A}, \mathbf{B} \in \mathbf{P(X)}$ be given. Then the fuzzy set $[A\,B]^{1/2}$ is the geometric mean of $\mathbf{A}$ and $\mathbf{B}$ on X, where for:

$$\mu_{[A \cdot B]^{1/2}}(x) = \left[\mu_A(x)\,\mu_B(x)\right]^{1/2}.$$

The weighting of the individual assessments and thus the willingness to compromise is defined in Definitions 2.27 and 2.28. The curve of the membership functions is located between the min and max operators.

Example 2.19 Figure 2.17 shows the curve of the membership functions of the arithmetic and geometric mean values for a given set according to Example 2.13.

An adjustable degree of compromise can be achieved, for example, by approximating the minimum/maximum operators to the arithmetic mean, as Werners (1984) proposed. This results in the operators *and* and *or*.

Definition 2.29 (Fuzzy *and*, Fuzzy *or*) Let $\mathbf{A_i} = \{(x, \mu_i(x)) \mid x \in X\}_{i=1,\dots,n}$ with $\mathbf{A_i} \in \mathbf{P(X)}$. Then the fuzzy set is expressed as *and* operator of $\mathbf{A_i}$ with:

$$\mu_{\mathrm{and}}(x) = \delta \cdot \min\left[\mu_1(x), \dots, \mu_n(x)\right] + (1 - \delta)/n \cdot \sum_i \mu_i(x),$$

the fuzzy set as *or* operator of $\mathbf{A_i}$ with:

$$\mu_{\mathrm{or}}(x) = \delta \cdot \max\left[\mu_1(x), \dots, \mu_n(x)\right] + (1 - \delta)/n \cdot \sum_i \mu_i(x).$$

The factor $(1 - \delta) \in [0, 1]$ expresses the orientation from the max/min value to the arithmetic mean.

Example 2.20 Let **A** and **B** be fuzzy sets according to Example 2.13. Then, for $\delta = 0.5$, the following applies:

$$\mathbf{A} \text{ and } \mathbf{B} = \{(5; 0.3),\ (6; 0.575),\ (7; 0.55),\ (8; 0.275),\ (9; 0)\},$$

$$\mathbf{A} \text{ or } \mathbf{B} = \{(5; 0.7),\ (6; 0.725),\ (7; 0.85),\ (8; 0.625),\ (9; 0)\}.$$

The operators *and* and *or* do not obey the laws for t-norm or t-conorm. The membership functions run between the arithmetic mean and the min or max combination. The degree of compensation for a specific *and* (and the corresponding *or*) is fixed with the parameter δ. To describe real decision-making situations, a large number of differently set *and/or* operators are therefore necessary.

Zimmermann and Zysno (1980) introduced a generalization of the operators presented so far: An operator is defined whose membership function can be variably set between the intersection and the union. For this purpose, the geometric mean is taken as the starting point. By rewriting Definition 2.28, the following initially follows:

$$[\mu_A(x) \cdot \mu_B(x)]^{1/2} = \mu_A(x)^{1/2} \cdot \mu_B(x)^{1/2}$$

$$= \min[\mu_A(x),\ \mu_B(x)]^{1/2} \cdot \max[\mu_A(x),\ \mu_B(x)]^{1/2}.$$

It is therefore clear that, in the geometric mean, the intersection and union are weighted equally with an exponent of 0.5. An imbalance arises through the introduction of a variable exponent $\gamma \in [0, 1]$, initially leading to the operator:

$$\min[\mu_A(x),\ \mu_B(x)]^{1-\gamma} \cdot \max[\mu_A(x),\ \mu_B(x)]^{\gamma}.$$

The exponent γ is called the "degree of compensation." Since the min/max combination neglects too much information, especially when more than two membership functions or evaluations are to be combined, they are replaced by the algebraic product and the algebraic sum to form the overall formulation of the "compensatory and."

Definition 2.30 (γ-Operation (Compensatory and)) The γ-operation of two fuzzy sets $\mathbf{A}, \mathbf{B} \in \mathbf{P(X)}$, written $\mathbf{A}\gamma\mathbf{B}$, produces a fuzzy set with the membership function:

$$\mu_{A\gamma B}(x) = \left[\mu_{A \cdot B}(x)\right]^{1-\gamma} \cdot \left[\mu_{A+B}(x)\right]^{\gamma} \tag{2.25}$$

and the degree of compensation $\gamma \in [0, 1]$. For example:

$$\begin{aligned}
\gamma = 0 \quad &\Rightarrow \mu_{A \cdot B}(x) &&[\text{algebraic product}], \\
\gamma = 0.5 \quad &\Rightarrow \left[\mu_{A \cdot B}(x) \cdot \mu_{A+B}(x)\right]^{0.5}, \\
\gamma = 1 \quad &\Rightarrow \mu_{A+B}(x) &&[\text{algebraic sum}].
\end{aligned}$$

The γ-operation can easily be extended to more than two sets to be combined. With the identity

$$u + v - uv = 1 - (1 - u)(1 - v), \qquad u, v \in \mathbb{R} \tag{2.26}$$

the second factor on the right-hand side of (2.25) can be represented as:

$$\mu_A(x) + \mu_B(x) - \mu_A(x) \cdot \mu_B(x) = 1 - \bigl(1 - \mu_A(x)\bigr)\bigl(1 - \mu_B(x)\bigr).$$

This form is suitable for generalizing (2.23) to multiple fuzzy sets to be combined.

Definition 2.31 (γ-Operation for Several Fuzzy Sets to be Combined) Let $\mathbf{A_1}, \ldots, \mathbf{A_n} \in \mathbf{P(X)}$. The γ-operation of the $\mathbf{A_i} = \{(x, \mu_i(x)) \mid x \in X\}$ leads to a fuzzy set A_γ with the membership function:

$$\mu_\gamma(x) = \left[\prod_i \mu_i(x)\right]^{1-\gamma} \cdot \left[1 - \prod_i (1 - \mu_i(x))\right]^\gamma$$

and $\gamma \in [0, 1]$ as the degree of compensation.

After this theoretical introduction, the question arises as to how to find an adequate degree of compensation for practical problems. The procedure described in Zimmermann and Zysno (1984) can be used for this purpose:

1. Conduct empirical preliminary tests with a representative subset $\Omega \subseteq X$.
2. Determine the $\mu_\gamma(x)$, $\forall x \in \Omega$ for each point, in addition to determining the $\mu_i(x)$.
3. Calculate a compensation function $\gamma(x)$, $x \in \Omega$.
4. Calculate an empirical compensation factor Γ by taking the arithmetic mean of $\gamma(x)$ over Ω.

Regarding 3: By taking the logarithm, we obtain the following for $\mu_i(x) \neq 0$:

$$\gamma(x) = \frac{\log \mu_\gamma(x) - \log\left[\prod_i \mu_i(x)\right]}{\log\left[1 - \prod_i (1 - \mu_i(x))\right] - \log\left[\prod_i \mu_i(x)\right]} \tag{2.27}$$

Regarding 4: The empirical compensation factor Γ is given by:

$$\Gamma = (1/|\Omega|) \cdot \sum_{x \in \Omega} \gamma(x). \tag{2.28}$$

With the help of (2.27) and (2.28), the non-compensatory operations can also be assessed in terms of their degree of compensation (i.e., their proximity to $(\cdot)$ or $(+)$). To do this, the respective operations' results are inserted selectively in (2.27) instead of $\mu_\gamma(x)$. The resulting ranking of the operations is generally valid, but the specific numerical values of the degrees of compensation vary with the data of the fuzzy input sets.

Table 2.1 Comparison of the
degrees of compensation of
different operators

Operator	Degree of compensation
$(+)$	1.00
(max)	0.96
$(A + B)/2$	0.67
$(A \cdot B)^{1/2}$	0.53
(min)	0.11
$(\cdot)$	0.00

Example 2.21 (Computer Production According to Example 2.13) The use of logarithms in (2.27) leads to the requirement $\mu_{A \cdot B}(x) \neq 0$, so that the support set $X' = \{5, 6, 7, 8\}$ is used as the base set for comparison. The following Table 2.1 can then be drawn up.

For further properties of fuzzy sets and operations with fuzzy sets (e.g., Bandemer, 1990 and Klir & Folger, 1988).

2.4 Measures of Uncertainty

Following the previous chapters, the question arises as to how the uncertainty of sets can be measured and thus compared with each other. In general, various measures of uncertainty are conceivable and applicable. What they all have in common is that they map the power set $\mathbf{P}(\mathbf{X})$ onto a metric scale (i.e., without restricting generality to the unit interval $[0, 1]$).

When defining a measure of uncertainty system, three specifications must be defined:

1. The reference set for the measure value 0 (i.e., the *crispest* possible set)
2. The reference set for the measure value 1 (i.e., the *fuzziest* possible set)
3. A comparison relation for the *fuzziness*

Two important measurement systems include entropy and energy measures. They are based on different premises in accordance with the above information and are presented below.

Entropy Measures as Measures of Uncertainty

All crisp sets are assigned a measure value of 0. The fuzziest set is the set whose membership function takes the constant value $\frac{1}{2}$ for all elements and thus runs in the middle of the unit interval $[0, 1]$. A comparison of uncertainty is made using the curve of the membership functions.

Definition 2.32 (Entropy Measure) For an entropy measure d of two fuzzy sets **A** and **B**, the following applies:

1. $d(A) = 0$ for $\mu_A : X \to \{0, 1\}$.
2. $d(A) = \max$ for $\mu_A : X \to \{\tfrac{1}{2}\}$.
3. $d(A) > d(B)$ for **B** steeper than **A**.

$$\left.\begin{array}{ll} \mu_B \leq \mu_A & \text{for } \mu_A < \tfrac{1}{2}, \\[2mm] \mu_B \geq \mu_A & \text{for } \mu_A > \tfrac{1}{2} \end{array}\right\} \quad \Longrightarrow \quad d(A) \geq d(B).$$

The uncertainty index (based on Yager, 1980) and *Shannon's entropy measure* are explained in more detail below.

Yager assumes the identity $\mathbf{A} \cap \mathbf{A^c} = \varnothing$ for *crisp* sets , which does not apply to fuzzy sets. First, the deviation $\mathbf{D} = \mathbf{A} \cap \mathbf{A^c}$ is calculated:

$$\begin{aligned} \mu_D(x) &= \min[\mu_A(x),\, 1 - \mu_A(x)] \\ &= \tfrac{1}{2}[\mu_A(x) + (1 - \mu_A(x)) - |\mu_A(x) - (1 - \mu_A(x))|] \\ &= \tfrac{1}{2}[1 - |\mu_A(x) - (1 - \mu_A(x))|]. \end{aligned}$$

To calculate the uncertainty index , the support set $\mathbf{S(D)}$ and the cardinality $|\mathbf{D}|$ are used:

$$|\mathbf{D}| = \sum_{x \in S(D)} \mu_D(x), \quad \forall x \in \mathbf{S(D)}. \tag{2.29}$$

Definition 2.33 (Uncertainty Index) Let $A \in P(X)$. Then

$$d_1(\mathbf{A}) = 1 - \rho_1(A, A^c)/|S(\mathbf{D})|$$

with

$$\begin{aligned} \rho_1(\mathbf{A}, \mathbf{A^c}) &= \sum_{x \in S(D)} |\mu_A(x) - (1 - \mu_A(x))| \\ &= \sum_{x \in S(D)} |2\mu_A(x) - 1|. \end{aligned}$$

The *uncertainty index* of A. The factor $\rho_1(\mathbf{A}, \mathbf{A^c})$ can be interpreted as the distance between the two sets A and A^c and is called the *Hamming distance* between **A** and $\mathbf{A^c}$.

Instead of $\rho_1(\mathbf{A}, \mathbf{A^c})$, one could also use the Euclidean distance:

$$\rho_1(\mathbf{A}, \mathbf{A^c}) = \left[\sum_{x \in S(D)} |\mu_A(x) - (1 - \mu_A(x))|^2\right]^{1/2}. \tag{2.30}$$

More generally, one can use a distance $\rho_p(A, A^c)$, which is calculated as follows:

$$\rho_p(\mathbf{A}, \mathbf{A}^c) = \left[\sum\nolimits_{x \in S(D)} |\mu_A(x) - (1 - \mu_A(x))|^p\right]^{1/p}. \tag{2.31}$$

Another way to define a measure of entropy is to use Shannon's concept of entropy with Shannon's entropy function $H(A)$. This results in a class of measures that depend on an adjustable parameter k.

Definition 2.34 (Shannon's Measure of Uncertainty) Let $\mathbf{A} \in \mathbf{P}(X)$. Then

$$d_2(\mathbf{A}) = -k \sum\nolimits_{x \in X} H(A) + H(A^c) \quad with$$

$$H(\mathbf{A}) = -\mu_A(x)\, \ln\big[\mu_A(x)\big].$$

Shannon's measure of uncertainty of $\mathbf{A}$. Here, k is a positive real number.

Theorem 2.5 *The following applies to the entropy measures described:*

$$d(\mathbf{A}) = d(\mathbf{A}^c). \tag{2.32}$$

Energy Measures as Measures of Uncertainty
Here, the reference set for the value 0 is the empty set $\varnothing$. The fuzziest set is considered to be the base set itself. A comparison is made using the criterion "containment."

Definition 2.35 (Energy Measure) For an energy measure r of two fuzzy quantities $\mathbf{A}$ and $\mathbf{B}$, the following applies:

1. $r(\mathbf{A}) = 0$ for $\mathbf{A} = \varnothing$.
2. $r(\mathbf{A}) = \max$ for $\mathbf{A} = \mathbf{E}$.
3. $\forall x \in X: \; \mu_A(x) \le \mu_B(x) \Rightarrow r(\mathbf{A}) \le r(\mathbf{B})$.

The following are examples of energy measures:

$$r_1(A) = |\mathbf{A}|. \tag{2.33}$$

$$r_2(A) = \mathrm{hgt}(\mathbf{A}) = \sup\nolimits_{x \in X} \mu_A(x). \tag{2.34}$$

$$r_3(A) = \int_X [\mu_A(x)]^2\, dx, \quad \text{if the integral exists.} \tag{2.35}$$

Given a specified entropy measure, a new energy measure can easily be generated and vice versa. According to Bandemer (1990), the following theorem applies:

Theorem 2.6 *Let $d_i(A)$ be an entropy measure for the fuzzy set $\mathbf{A}$. Furthermore, let $\mathbf{U}$ be a fuzzy set with:*

$$\mathbf{U} = \left\{ (x; \mu_{1/2}(x)) \mid \mu_{1/2}(x) = \tfrac{1}{2} \ \forall x \in X \right\}.$$

Then

$$r(\mathbf{A}) = d(\mathbf{A} \cap \mathbf{U}) \tag{2.36}$$

is an energy measure. If $r_j(\mathbf{A})$ is an energy measure, then conversely

$$d(A) = r(\mathbf{A} \cap \mathbf{A}^C) \tag{2.37}$$

is an entropy measure.

References

Bandemer, H. (1990). *Fuzzy data analysis*. Springer.

Bellman, R. E., & Giertz, M. (1978). On the analytic formalism of the theory of fuzzy sets. *Information Sciences, 5*(2), 149–156. https://doi.org/10.1016/0020-0255(73)90033-2

Bocklisch, F. (1988). *Prozessanalyse mit unscharfen Verfahren*. VEB Verlag Technik. Berlin.

Bothe, H.-H. (1995). *Bewertung mit unscharfen Mengen*. Technische Hochschule Ilmenau.

Klir, G. J., & Folger, T. A. (1988). *Fuzzy sets, uncertainty, and information*. Prentice Hall.

Rommelfanger, H. (1988). *Fuzzy decision support systems*. Springer.

Werners, B. (1984). A typology of aggregation operators in mcdm. *Fuzzy Sets and Systems, 13*(1), 197–201. https://doi.org/10.1016/0165-0114(84)90048-4

Yager, R. R. (1980). A measure of uncertainty in fuzzy sets. *Information Sciences, 20*(2), 171–178. https://doi.org/10.1016/0020-0255(80)90065-8

Zadeh. (1965). Fuzzy sets. *Information and Control, 8*, 338–353. https://doi.org/10.1016/S0019-9958(65)90241-X

Zadeh. (1973). Outline of a new approach to the analysis of complex systems and decision processes. *IEEE Transactions on Systems, Man, and Cybernetics, SMC-3*(1), 28–44. https://doi.org/10.1109/TSMC.1973.5408575

Zimmermann, H.-J. (1991). *Fuzzy set theory – and its applications* (2nd ed.). Kluwer Academic Publishers.

Zimmermann, H.-J., & Zysno, P. (1980). Latent connectives in human decision making. *Fuzzy Sets and Systems, 3*(4), 287–305. https://doi.org/10.1016/0165-0114(80)90018-5

Zimmermann, H.-J., & Zysno, P. (1984). Decision and evaluation by multiple-attribute utility theory: The role of semantics in fuzzy data analysis. *Fuzzy Sets and Systems, 14*(2), 175–193. https://doi.org/10.1016/0165-0114(84)90095-2

Chapter 3
Fuzzy Numbers and Arithmetic

Abstract This chapter presents fuzzy numbers as extensions of real numbers that represent uncertain or imprecise values, rather than single points. Fuzzy arithmetic, to which the second part is devoted, involves performing addition, subtraction, multiplication, and division on these fuzzy numbers.

Fuzzy numbers are fuzzy sets with special properties. According to Bothe (1995), they play a role, for example, as possible values of the input or output variables of a system. In Sect. 3.1, we will therefore first clarify some terms before going into more detail on the calculation rules in Sect. 3.2.

3.1 Fuzzy Numbers

Definition 3.1 (Fuzzy Number) A convex, normalized fuzzy set $\mathbf{A}$ on the set $\mathbb{R}$ of real numbers is called a fuzzy number if:

1. There exists exactly one real number a with $\mu_A(a) = 1$.
2. $\mu_A(x)$ is at least piecewise continuous.

Point a is called the mean value of $\mathbf{A}$. A non-empty fuzzy number is called positive, if $\forall x \leq 0 : \mu_A(x) = 0$. This is written as $\mathbf{A} > 0$. A non-empty fuzzy number is called negative, if $\forall x > 0 : \mu_A(x) = 0$. This is written as $\mathbf{A} < 0$. A fuzzy number with a completely continuous membership function is called a continuous fuzzy number.

Example 3.1 The fuzzy set $\mathbf{A}$ on $\mathbb{R}$ with the membership function

$$\mu_A(x) = \begin{cases} (1+x)/4 & & x \in [-1, 3], \\ (5-x)/2 & \text{for} & x \in [3, 5], \\ 0 & & \text{otherwise} \end{cases}$$

H.-H. Bothe, E. Portmann, *Computing with Words*, Fuzzy Management Methods,
https://doi.org/10.1007/978-3-032-24117-7_3

Fig. 3.1 Membership function for a fuzzy number *approximately equal to 3*

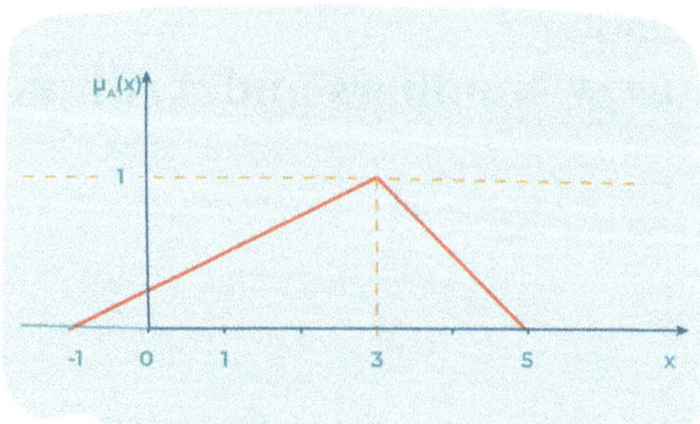

Fig. 3.2 Membership functions for three fuzzy sets **A**, **B**, and **C**

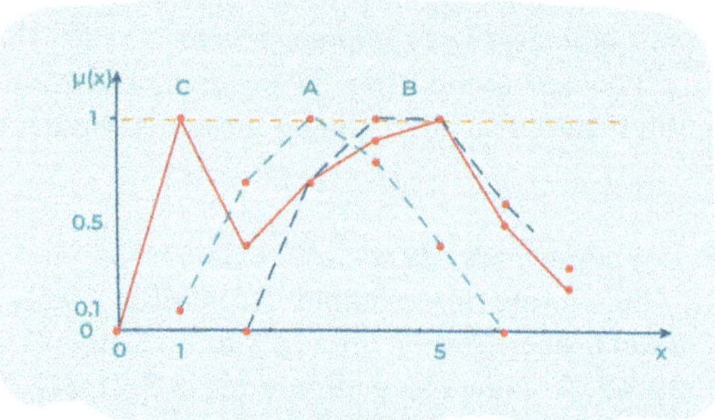

is a fuzzy number *approximately equal to 3*. It is neither positive nor negative (Fig. 3.1).

Definition 3.2 (Discrete Fuzzy Number) A fuzzy set **N** on a countable base set $X \subset \mathbb{R}$ is called a discrete fuzzy number if there exists a fuzzy number **A** such that $\forall x \in X$:

$$\mu_N(x) = \mu_A(x).$$

A simple way to generate **A**, given a fuzzy set **N** on a finite base set, is to connect all points $(x, \mu_N(x))$ using polygonal lines. The continuity condition (2) in Definition 3.1 is thus satisfied. The following example illustrates Definition 3.2.

Example 3.2 Let the following fuzzy sets **A**, **B**, and **C** be given:
$A = \{(1, 0.1), (2, 0.7), (3, 1), (4, 0.8), (5, 0.4), (6, 0)\}$,
$B = \{(2, 0), (3, 0.7), (4, 1), (5, 1), (6, 0.6), (7, 0.3)\}$,
$C = \{(0, 0), (1, 1), (2, 0.4), (3, 0.7), (4, 0.9), (5, 1), (6, 0.5), (7, 0.2)\}$,
and the resulting polygonal lines are plotted in Fig. 3.2.

According to Definition 3.2, only **A** is a discrete fuzzy number, whereas **C** is not convex and there is no unique vertex for **B**.

Definition 3.3 (Fuzzy Interval) A convex, normalized fuzzy set A on $\mathbb{R}$ is called a fuzzy interval if:

1. There exist two real numbers m_1, m_2 such that $\forall x \in [m_1, m_2] : \mu_A(x) = 1$.
2. $\mu_A(x)$ is piecewise continuous.

Note In Example 3.2, B is a fuzzy interval. In addition to the distinction made here between fuzzy numbers and intervals, the first condition of Definition 3.1 is sometimes modified so that fuzzy intervals are also considered as fuzzy numbers.

3.2 Extension Principle

To derive calculation rules between fuzzy numbers, the "extension principle" presented in its first version by Zadeh (1965, 1973) is used. This creates a tool for transferring mathematical concepts for crisp sets as well as numbers to fuzzy sets as well as numbers. This makes it possible to define mappings in the form of functions and relations between fuzzy numbers and sets. For this purpose, the term "Cartesian product" is first extended to fuzzy sets.

Definition 3.4 (Cartesian Product Between Fuzzy Sets) Let $A_1, \ldots, A_n$ be fuzzy sets on $X_1, \ldots, X_n$. Then the Cartesian product $A_1 \otimes \cdots \otimes A_n$ is a fuzzy set in the product space $X = X_1 \times \cdots \times X_n$ with:

$$\mu_{kp}(x_1, \ldots, x_n) = \min \left[\mu_{A_i}(x_i) \mid i = 1, \ldots, n, \ x_i \in X_i \right].$$

Example 3.3 (Cartesian Product) Let:
$A_1 = \{(4, 0.4), (5, 0.7), (6, 1), (7, 0.4)\}$ and
$A_2 = \{(2, 0.1), (3, 0.6), (4, 1), (5, 0.7), (6, 0.3)\}$.
Then $\mu_{kp}(x_1, x_2)$ is the two-dimensional membership matrix of the Cartesian product $A_1 \otimes A_2$:

$\mu_{kp}(x_1, x_2): x_1 \backslash x_2$	2	3	4	5	6
4	0.1	0.4	0.4	0.4	0.3
5	0.1	0.6	0.7	0.7	0.3
6	0.1	0.6	1	0.7	0.3
7	0.1	0.4	0.4	0.4	0.3

The use of the min operator in the definition of the Cartesian product means that the resulting value tuples with the smallest membership value belong to the product space. Based on the Cartesian product, a general rule for defining calculation rules in the area of fuzzy sets or numbers can now be defined.

Definition 3.5 (Extension Principle (Eigenvalue Problem))   *Given*:

1. $n + 1$ base sets $X_1, \ldots, X_n, Y$.
2. n fuzzy sets $\mathbf{A_i}$ on X_i with membership functions $\mu_i(x)$, $i = 1, \ldots, n$.
3. a mapping $f : X_1 \times \cdots \times X_n \to Y$ with $y = f(x_1, \ldots, x_n)$ and $y \in Y$.

The mapping $f(\mathbf{A_1}, \ldots, \mathbf{A_n})$ leads to a fuzzy set $\mathbf{B}$ on Y with:

$$B = \{(y; \mu_B(y)) \mid y = f(x_1, \ldots, x_n),\ (x_1, \ldots, x_n) \in X_1 \times \cdots \times X_n\},$$

where the membership function $\mu_B(y)$ is given by:

$$\mu_B(y) = \begin{cases} \sup_{y = f(x_1, \ldots, x_n)} \min\big[\mu_1(x_1), \ldots, \mu_n(x_n)\big], & \text{if } \exists\, y = f(x_1, \ldots, x_n), \\ 0, & \text{otherwise.} \end{cases}$$

According to Definition 3.5, $\mu_B(y)$ is calculated from the supremum for $\mu_{kp}(x_1, \ldots, x_n)$ with the n-tuples $(x_1, \ldots, x_n)$ leading to the result value $y = f(x_1, \ldots, x_n)$.

For the special case $n = 1$, the calculation of the Cartesian product is omitted, and the eigenvalue problem simplifies to:

$$\mathbf{B} = f(\mathbf{A}) = \{(y; \mu_B(y)) \mid y = f(x),\ x \in X\} \quad with$$

$$\mu_B(y) = \begin{cases} \sup_{y = f(x)} \mu_A(x) & \text{if } y = f(x) \text{ exists,} \\ 0 & \text{otherwise.} \end{cases} \tag{3.1}$$

Example 3.4 (One-Dimensional Quadratic Function)   Let:

$$f(x) = x^2 + 1 \quad \text{and} \quad \mathbf{A} = \{(-1; 0.5),\ (0; 0.08),\ (1; 1),\ (2; 0.4)\}.$$

According to (3.1), the result set $\mathbf{B} = f(\mathbf{A})$ is calculated as:

$$\mathbf{B} = \{(1; 0.08),\ (2; 1),\ (5; 0.4)\}.$$

The mapping $\mathbf{B} = f(\mathbf{A})$ can be represented graphically as follows, where $S(\mathbf{A})$ and $S(\mathbf{B})$ are the support sets of $\mathbf{A}$ and $\mathbf{B}$ (Fig. 3.3):

Example 3.5 (Extended Addition of Two Fuzzy Sets)   Let the fuzzy sets $\mathbf{A_1}$ and $\mathbf{A_2}$ (corresponding to Example 3.3) be given with:

$$\mathbf{A_1} = \{(4; 0.4),\ (5; 0.7),\ (6; 1),\ (7; 0.4)\} \text{ and}$$

$$\mathbf{A_2} = \{(2; 0.1),\ (3; 0.6),\ (4; 1),\ (5; 0.7),\ (6; 0.3)\},$$

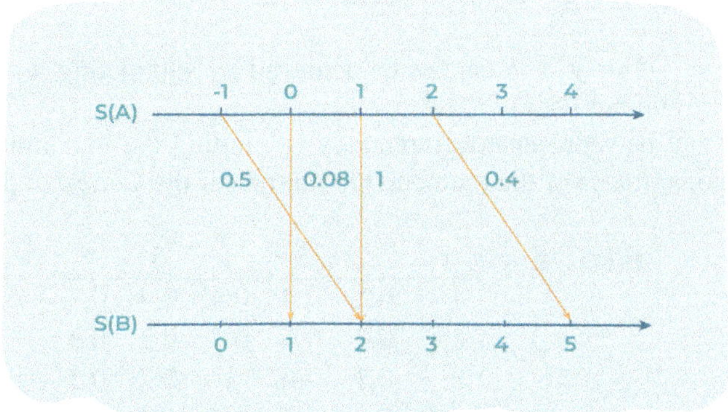

Fig. 3.3 Representation of the mapping $\mathbf{B} = f(\mathbf{A})$ with $f(x) = x^2 + 1$

and the function

$$f(x_1, x_2) = x_1 + x_2 \text{ with } x_1 \in \mathbf{A_1} \text{ and } x_2 \in \mathbf{A_2}.$$

The fuzzy result set $\mathbf{B} = f(\mathbf{A_1}, \mathbf{A_2})$ is calculated using the Cartesian product (according to Example 3.3) to:

$$B = \{(6; 0.1),\ (7; 0.4),\ (8; 0.6),\ (9; 0.7),\ (10; 1),\ (11; 0.7),\ (12; 0.4),\ (13; 0.3)\}.$$

The values $y < 6$ or $y > 13$ cannot be achieved by adding $\mathbf{A}_1$ and $\mathbf{A}_2$.

The extended addition is also written as $\mathbf{B} = \mathbf{A_1} \oplus \mathbf{A_2}$. The result can be represented graphically by finding the maximum within the indicated diagonal traces of the membership matrix for the Cartesian product:

$\mu_{kp}(x_1, x_2) : x_1 \backslash x_2$	2	3	4	5	6
4	0.1	0.4	0.4	0.4	0.3
5	0.1	0.6	0.7	0.7	0.3
6	0.1	0.6	1	0.7	0.3
7	0.1	0.4	0.4	0.4	0.3

Example 3.6 (Extended Subtraction of Two Fuzzy Sets) Let there be two fuzzy sets $\mathbf{A_1}$ and $\mathbf{A_2}$ (as in Example 3.5), as well as the function:

$$f(x_1, x_2) = x_1 - x_2 \text{ with } x_1 \in \mathbf{A_1} \text{ and } x_2 \in \mathbf{A_2}.$$

The fuzzy result set $\mathbf{B} = f(\mathbf{A_1}, \mathbf{A_2}) = \mathbf{A_1} \ominus \mathbf{A_2}$ describes a subtraction (≈ 6)–(≈ 4) and is calculated using the Cartesian product (according to Example 3.5) to:

$\mathbf{B} = \{(-2; 0.3),\ (-1; 0.4),\ (0; 0.7),\ (1; 0.7),\ (2; 1),\ (3; 0.6),\ (4; 0.4),\ (5; 0.1)\}.$

The values $y < -2$ or $y > 5$ cannot be achieved by subtracting $\mathbf{A_1}$ and $\mathbf{A_2}$. The result is a fuzzy number (≈ 2).

The result can be represented graphically by finding the maximum within the indicated diagonal traces of the membership matrix for the Cartesian product:

$\mu_{kp}(x_1, x_2) : x_1 \backslash x_2$	2	3	4	5	6
4	0.1	0.4	0.4	0.4	0.3
5	0.1	0.6	0.7	0.7	0.3
6	0.1	0.6	1	0.7	0.3
7	0.1	0.4	0.4	0.4	0.3

The extended multiplication and division of fuzzy numbers can be calculated using the same method as in the two previous examples above. The basic arithmetic operators extended to the domain of fuzzy numbers are written in the following notation: "$\oplus, \ominus, \odot, \oslash$."[1] The question remains open as to whether or when the result of the calculation is again a fuzzy number. The investigation can be generalized for different operators. Let $\square$ be a binary operation on $\mathbb{R}$ that is extended to an operation $\blacksquare$ for the domain of fuzzy numbers. According to the eigenvalue problem , the operation $\mathbf{A} \blacksquare \mathbf{B}$ results in a fuzzy set with the membership function:

$$\mu_{A \blacksquare B}(y) = \sup_{y = x_1 \square x_2} \left\{ \min\left[\mu_A(x_1), \mu_B(x_2) \right] \right\}. \tag{3.2}$$

Theorem 3.1

1. *For every commutative operation $\square$, $\blacksquare$ is also commutative.*
2. *For every associative operation $\square$, $\blacksquare$ is also associative.*[2]

The following definition serves to simplify the formulation of the two theorems that follow.

Definition 3.6 (Monotonicity)   A binary operation $\square$ on a basic domain[3] $\mathbb{R}$, $\mathbb{R}^+$ bzw. $\mathbb{R}^-$ is called strictly monotonically increasing (or decreasing) if the following holds for $x_1 > y_1$ and $x_2 > y_2$:

$$x_1 \square x_2 > y_1 \square y_2 \quad (\text{or } x_1 \square x_2 < y_1 \square y_2).$$

---

[1] An extension to more than binary operators is discussed, for example, in Dubois and Prade (1979).

[2] If associativity is valid, recursive calculation algorithms can also be defined.

[3] The basic domain for the operation is specified in one dimension because it is executed between two operators with the same basic domain.

Example 3.7

$$f(x_1, x_2) = x_1 + x_2 \quad \text{is strictly monotonically increasing on } \mathbb{R}.$$
$$f(x_1, x_2) = x_1 \cdot x_2 \quad \text{is strictly monotonically increasing on } \mathbb{R}^+.$$
$$f(x_1, x_2) = -(x_1 + x_2) \text{ is strictly monotonically decreasing on } \mathbb{R}.$$
$$f(x_1, x_2) = x_1/x_2 \quad \text{is neither strictly monotonically increasing nor monotonically decreasing on } \mathbb{R}.$$

Theorem 3.2 *Let **A** and **B** be the two continuous fuzzy numbers and $\square$ the continuous,strictly monotonic operation on $\mathbb{R}$. Then the result **A** $\blacksquare$ **B** also forms a continuous fuzzy number.*

Theorem 3.3 (e.g., Rommelfanger (1988)) can serve as a practical basis for calculating the result of the operation. The method consists of dividing the membership functions $\mu_A(x)$ and $\mu_B(x)$ into a monotonically increasing and a monotonically decreasing branch, respectively, and treating the increasing and decreasing branches of A and B together. If plateaus occur, they are processed in a joint calculation step.

Theorem 3.3 *Let **A** and **B** be the two continuous fuzzy numbers and $\square$ the continuous, strictly monotonic operation. Let $[a_1, a_2]$ and $[b_1, b_2]$ be the intervals in which $\mu_A(x)$ and $\mu_B(x)$ rise (or fall) monotonically. If subintervals $[\alpha_1, \alpha_2] \subseteq [a_1, a_2]$ and $[\beta_1, \beta_2] \subseteq [b_1, b_2]$ exist with:*

$$\mu_A(x_A) = \mu_B(x_B) = \lambda \quad \forall x_A \in [\alpha_1, \alpha_2], \ x_B \in [\beta_1, \beta_2],$$
$$\text{it holds that: } \mu_{A \blacksquare B}(t) = \lambda \quad \forall t \in [\alpha_1 \ \square \ \beta_1, \ \alpha_2 \ \square \ \beta_2].$$

Remarks

1. If $(\alpha_1 = \alpha_2) \vee (\beta_1 = \beta_2)$, then the operation is performed pointwise in the subinterval of the branch with plateau.
2. If $(\alpha_1 = \alpha_2) \wedge (\beta_1 = \beta_2)$ applies, stepping through the respective branches of **A** and **B** in the value range $[0, 1]$ exactly at the point $t = x_A \ \square \ x_B$ leads to the membership value $\mu_{A \blacksquare B}(t) = \lambda$.
3. Theorem 3.2 can be interpreted as follows: The operation to be extended is used to calculate points with a specific membership value of the extended operation.

Example 3.8 (Further Extended Addition of Two Fuzzy Numbers) Let the two fuzzy numbers **A** and **B** be given as shown in Fig. 3.4, which are to be added according to Theorem 3.3. The result is the fuzzy number **C**.

According to Theorem 3.3, for example, the following applies to the left subset of **C**

$$\mu_{A \blacksquare B}(t_L) = 0.4 \quad \forall t_L \in [20 + 64, \ 30 + 64],$$

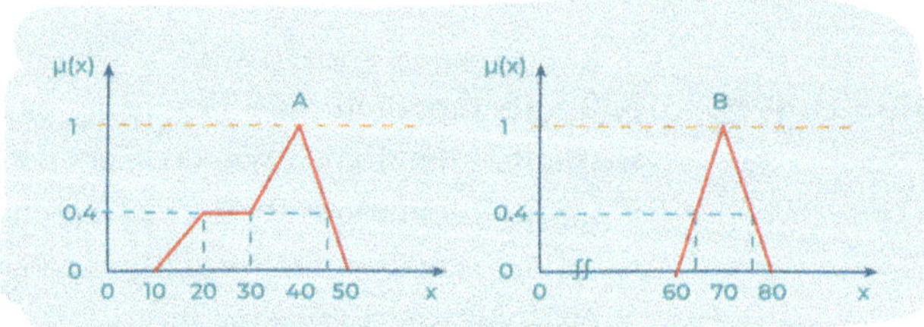

Fig. 3.4 Membership functions of two fuzzy numbers **A** and **B** that are to be added

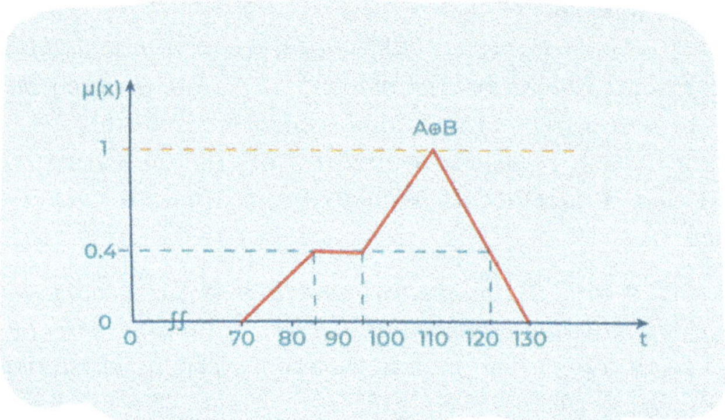

Fig. 3.5 Membership function of $\mathbf{C} = \mathbf{A} \oplus \mathbf{B}$

for the right subset of **C**, this results in a membership value of 0.4 at the point $t_R = 46 + 76 = 122$. The mean value of **C** is at the point $c_0 = 40 + 70 = 110$. The support set of **C** is $S(\mathbf{C}) = [10 + 60,\ 50 + 80] = [70,\ 130]$.

The extended addition of **A** and **B** thus leads to the following fuzzy number **C** (Fig. 3.5):

Example 3.9 Let **A** and **B** be the fuzzy numbers on $\mathbf{P}(X)$ (as shown in Fig. 3.6). Then $\max(\mathbf{A}, \mathbf{B})$ is a strictly monotonically increasing function on $\mathbb{R}$. According to Theorem 3.2, $\mathbf{C} = \max(\mathbf{A}, \mathbf{B})$ is also a fuzzy number. According to Theorem 3.3, the membership function for **C** is illustrated in Fig. 3.6.

An extension of the calculation method according to Theorem 3.3 to arbitrary continuous fuzzy sets can be performed by decomposing them into intervals with strictly monotonic or constant behavior. However, this results in a relatively large amount of numerical computation.

In addition to the extension principle specified in Definition 3.5, there are other proposals for transferring mapping rules from conventional mathematics to the domain of fuzzy sets or numbers, for example:

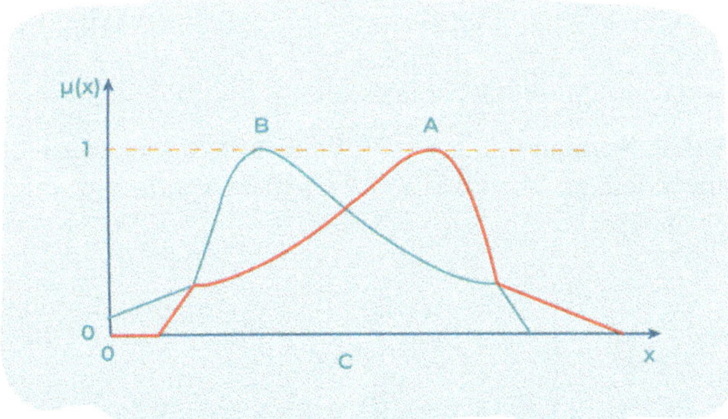

Fig. 3.6 Extended formation of the maximum of two fuzzy numbers **A** and **B**

1. Replacement of the logical "and" (min) in the Cartesian product by the algebraic product or the γ-operator.
2. Replacing the supremum by the algebraic sum .

3.3 LR Numbers and Operations

Calculating with fuzzy numbers can be simplified by requiring that only functions of a certain, predetermined type are allowed for the two branches of the membership function to the left and right of the mean value. To this end, requirements are placed on the curve of the branches to ensure that the fuzzy sets represented are indeed fuzzy numbers. This results in a selection of possible function curves. This produces fuzzy numbers that can be assigned to a specific type. If these are linked together, the calculation results can also be represented in the same scheme. Under certain circumstances, however, an approximation of the actual result may be necessary in order to obtain a fuzzy number of the same type. A relevant process function for the branches of fuzzy numbers is called a shape function; the fuzzy numbers themselves are called LR numbers in this representation. Very often, the shape functions are composed of straight-line segments, resulting in LR numbers of the triangle type. The exact definitions are given below.

Definition 3.7 (Shape Function) A function $S : [0, \infty) \to [0, 1]$ is called a shape function if:

1. $S(0) = 1$.
2. S is monotonically decreasing in the range $[0, \infty)$.

Common examples are the following parametric shape functions, where the argument $u = u(x)$ is a function of the elements x of the basic domain X of the fuzzy number:

$$S_1(u) = (1 + u^p)^{-1}, \quad S_2(u) = e^{-u^p}, \quad S_3(u) = \max(0, \, 1 - u^p),$$

where $p > 0$ is a constant setting parameter.

Definition 3.8 (LR Number) Let $\mathbf{A} \in \mathbf{P}(X)$ with $\mu_A(a) = 1$ and $L(u)$ and $R(u)$ be the two shape functions. Then $\mathbf{A}$ is an *LR number* (or a fuzzy number of type LR) if the following applies:

$$\mu_A(x) = \begin{cases} L((a - x)/\alpha) & x \le a, \; \alpha > 0, \\ & \text{for} \\ R((x - a)/\beta) & x \ge a, \; \beta > 0. \end{cases}$$

The coefficient a is called the mean value of $\mathbf{A}$ with $\mu_A(a) = 1$. $L(0) = R(0) = 1$, and α and β are called the left and right ranges of $\mathbf{A}$. The case $\alpha = \beta = 0$ characterizes the crisp number a. As the ranges increase, $\mathbf{A}$ becomes fuzzier. The following abbreviated notation is introduced for LR numbers:

$$\mathbf{A} = \langle a; \, \alpha; \, \beta \rangle_{\text{LR}}.$$

Example 3.10 Let $\mathbf{A}$ be the fuzzy number *approximately equal to 5* with the left and right shape functions:

$$L(u) = \max[0, 1 - u] \quad \text{and} \quad R(u) = \left[1 + u^2\right]^{-1}.$$

According to Definition 3.8, this means for the left branch of the membership function:

$$L(u(x)) = \max\left[0, 1 - (a - x)/\alpha\right]$$
$$= \max\left[0, 1/\alpha \cdot x + (\alpha - a)/\alpha\right],$$

and for the right branch:

$$R(u(x)) = \left[1 + ((x - a)/\beta)^2\right]^{-1}.$$

$L(u)$ is obviously a straight line in a segment with a positive slope $\Delta\mu/\Delta x = 1/\alpha$. The term $\max[0, \cdot]$ limits $L(u)$ to positive values. The resulting inflection point x_L of $L(u(x))$ is calculated as:

$$x_L = a - \alpha, \quad \text{where } L(u(x_L)) = 0.$$

For the special case $\mathbf{A} = \langle 5; 2; 1 \rangle_{LR}$, the following applies to the left branch of the membership function:

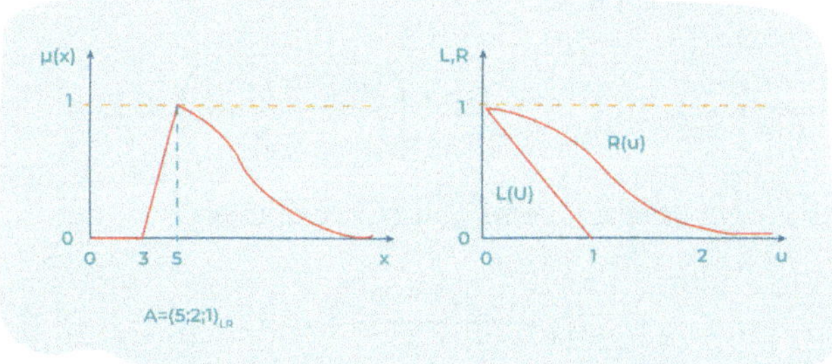

Fig. 3.7 Membership function (left) and shape functions (right) of the fuzzy number *approximately equal to 5*

$$L\big(u(x)\big) = \max\big[0,\ x/2 - 3/2\big]$$

and for the right branch:

$$R\big(u(x)\big) = \big[1 + [(x - 5)]^2\big]^{-1}.$$

The membership function and shape functions can be represented graphically (as shown in Fig. 3.7).

For LR numbers, the basic arithmetic operations according to the eigenvalue problem mentioned in Definition 3.5 will now be examined in more detail. A distinction must be made between whether the fuzzy numbers to be combined are of the same type or not and whether they are positive, negative, or neither positive nor negative. The most important cases are presented, but fuzzy numbers that are neither positive nor negative are not considered.

3.3.1   Extended Addition

Same LR type: $M = \langle m;\ \alpha;\ \beta \rangle_{LR}$, $\qquad N = \langle n;\ \gamma;\ \delta \rangle_{LR}$
If, for every fixed value $\lambda \in [0, 1]$, there are uniquely determined real numbers x and y with:

$$L[(m - x)/\alpha] = \lambda = L\lfloor(n - y)/\gamma\rfloor, \tag{3.3a}$$

then, according to Theorem 3.3, the following applies:

$$x = m - \alpha\, L^{-1}(\lambda), \quad y = n - \gamma\, L^{-1}(\lambda) \quad and$$

$$z = x = y = m + n - (\alpha + \gamma)L^{-1}(\lambda). \tag{3.3b}$$

Therefore, the following applies:

$$\frac{(m+n)-z}{\alpha+\gamma} = L^{-1}(\lambda) \;\Rightarrow\; L\left(\frac{(m+n)-z}{\alpha+\gamma}\right) = \lambda. \tag{3.3c}$$

Similarly, for the right branches of M and N, we obtain:

$$R\left(\frac{z-(m+n)}{\beta+\delta}\right) = \lambda. \tag{3.3d}$$

This implies the following for the sum result:

$$\langle m; \alpha; \beta\rangle_{LR} \;\oplus\; \langle n; \gamma; \delta\rangle_{LR} = \langle m+n; \; \alpha+\gamma; \; \beta+\delta\rangle_{LR}. \tag{3.4}$$

By substituting $u_L = (m-x)/\alpha$ and $u_R = (x-m)/\beta$, the dependence on x is eliminated, so that the extremely simple symbolic calculation method (3.4) can be applied. This applies to any reference functions that satisfy condition (3.3a).

Example 3.11 Let $M = \langle 5; 3; 1\rangle_{LR}$ and $N = \langle 3; 1; 2\rangle_{LR}$ be the representatives of fuzzy numbers *approximately equal to 5* and *approximately equal to 3*. Then $M \oplus N = \langle 8; 4; 3\rangle_{LR}$.
Representation for the case $L(u) = \max[0,\, 1-u]$, $\quad R(u) = \max[0,\, 1-u^2]$:
Different $\;M = \langle m; \alpha; \beta\rangle_{LR}$, $\;N = \langle n; \gamma; \delta\rangle_{L'R'}$.
An analogous proof as in the above case leads to:

$$\langle m; \alpha; \beta\rangle_{LR} \;\oplus\; \langle n; \gamma; \delta\rangle_{L'R'} = \langle m+n; \; 1; \; 1\rangle_{L''R''}, \tag{3.5a}$$

with the two shape functions deviating from the original functions $L(u)$ and $R(u)$:

$$L'' = \left(\alpha L^{-1} + \gamma L^{-1}\right)^{-1} \quad \text{and} \quad R'' = \left(\beta R^{-1} + \delta R^{-1}\right)^{-1}. \tag{3.5b}$$

3.3.2　Extended Negation: $\ominus\,N$

According to the extension law (or Theorem 3.3), the following applies: $\mu_{\ominus N}(x) = \mu_N(-x)\;\;\forall x \in \mathbb{R}$. The negation of a fuzzy number N is therefore performed by inverting the mean value $n \to -n$ and then swapping the two branches $R(u)$ and $L(u)$. An originally positive number of type LR becomes a negative number of type RL. The following applies:

$$\ominus \langle n; \alpha; \beta\rangle_{LR} = \langle -n; \beta; \alpha\rangle_{RL}. \tag{3.6}$$

3.3.3 *Extended Subtraction:* $\mathbf{M} \ominus \mathbf{N}$ *(with* $\mathbf{M}, \mathbf{N} > 0$*)*

Subtraction can be understood as the addition of a negative number. The following applies as an example:

$$
\begin{aligned}
\langle m; \alpha; \beta \rangle_{\text{LR}} \ominus \langle n; \gamma; \delta \rangle_{\text{RL}} &= \langle m; \alpha; \beta \rangle_{\text{LR}} \oplus \langle -n; \delta; \gamma \rangle_{\text{LR}} \\
&= \langle m - n; \ \alpha + \delta; \ \beta + \gamma \rangle_{\text{LR}}.
\end{aligned}
\tag{3.7}
$$

In (3.7), it is important to note that the numbers to be added must be of the same type (i.e., if the minuend is of type LR, the subtrahend must be of type RL).

Example 3.12 Let $\mathbf{M} = \langle 5; \ 3; \ 1 \rangle_{\text{LR}}$ and $\mathbf{N} = \langle 4; \ 1; \ 1 \rangle_{\text{RL}}$. Then:

$$
\mathbf{M} \ominus \mathbf{N} = \langle 1; \ 4; \ 2 \rangle_{\text{LR}}.
$$

Representation for the case $L(u) = R(u) = \max[0, \ 1 - u]$:

3.3.4 *Extended Multiplication*

Special cases with positive or negative fuzzy numbers of the same LR type should be handled. If the multiplicand or multiplier is neither positive nor negative, the following applies analogously to the corresponding branches. However, this results in considerably more computing effort.

Extended Multiplication of a Fuzzy Number by a Scalar

The representation of the result depends on the sign of the scalar. In the case of $\lambda < 0$, the two branches of the result are swapped in the same way as in (3.7). The following applies:

$$
\lambda > 0 : \ \lambda \odot \langle m; \alpha; \beta \rangle_{\text{LR}} = \langle \lambda m; \ \lambda \alpha; \ \lambda \beta \rangle_{\text{LR}},
\tag{3.8a}
$$

$$
\lambda < 0 : \ \lambda \odot \langle m; \alpha; \beta \rangle_{\text{LR}} = \langle \lambda m; \ -\lambda \beta; \ -\lambda \alpha \rangle_{\text{RL}}.
\tag{3.8b}
$$

Extended Multiplication of Two Fuzzy Numbers $\mathbf{M}, \mathbf{N} > 0$

According to (3.3b), the following applies:

$$
z = x \cdot y = m \cdot n - (m\gamma + n\alpha) \, L^{-1}(h) + \alpha\gamma \, [L^{-1}(h)]^2.
\tag{3.9}
$$

Thus, the resolution according to $L^{-1}(h)$ does not generally lead to the specified LR type for $\mathbf{M} \odot \mathbf{N}$. However, approximation formulas can be specified. For example, this results in:

1. $\alpha\gamma < m\gamma + n\alpha$ (quadratic expression in (3.9) neglected):

$$\Rightarrow \mathbf{M} \odot \mathbf{N} \approx \langle mn;\ m\gamma + n\alpha;\ m\delta + n\beta \rangle_{\mathrm{LR}}. \tag{3.10}$$

2. $\alpha\gamma \not< m\gamma + n\alpha$ (quadratic expression in (3.9) replaced by linear):

$$\Rightarrow \mathbf{M} \odot \mathbf{N} \approx \langle mn;\ m\gamma + n\alpha - \alpha\gamma;\ m\delta + n\beta + \beta\gamma \rangle_{\mathrm{LR}}. \tag{3.11}$$

The approximation according to (3.10) reflects the function curve near the mean value. The approximation according to (3.11) is based on a suggestion by Dubois and Prade (1980). If the quadratic term in (3.9) is replaced by a linear term, (3.9) can be clearly resolved to L^{-1}. The approximated membership function of the LR type then matches the actual curve of $\mu_{M \odot N}(x)$ at the following three points (which can be easily verified):

$$P_1 = [mn;\ 1], \quad P_2 = \big[(m-\alpha)(n-\gamma);\ L(1)\big], \quad P_3 = \big[(m+\beta)(n+\delta);\ R(1)\big].$$

Example 3.13 Let the fuzzy numbers $\mathbf{M} = \langle 3;\ 1;\ 2\rangle_{\mathrm{LR}}$ and $\mathbf{N} = \langle 4;\ 2;\ 1\rangle_{\mathrm{LR}}$ be given with the two reference functions $L(u) = R(u) = \max[0, 1 - u]$. As approximations for $\mathbf{M} \odot \mathbf{N}$ with the requirement of the same LR type, we obtain (3.10) and (3.11):

$$[\mathbf{M} \odot \mathbf{N}]_1 = \langle 12;\ 10;\ 11\rangle_{\mathrm{LR}}, \quad [\mathbf{M} \odot \mathbf{N}]_2 = \langle 12;\ 8;\ 13\rangle_{\mathrm{LR}}.$$

The exact and the two approximated membership curves are shown in Fig. 3.8 (Figs. 3.9 and 3.10).

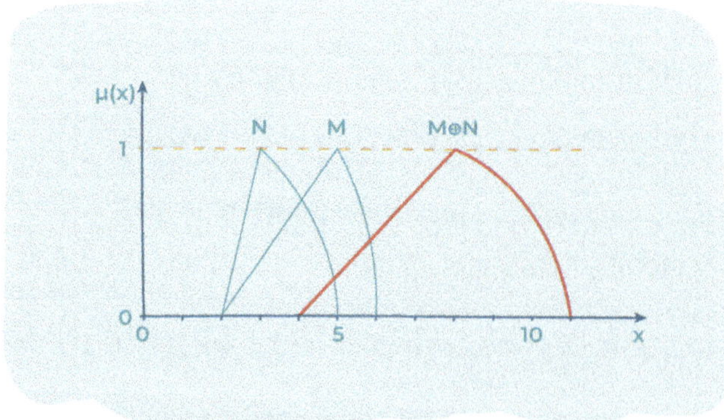

Fig. 3.8 Addition of two fuzzy numbers $\mathbf{M}$ and $\mathbf{N}$ ($\mathbf{M} \oplus \mathbf{N}$, bold)

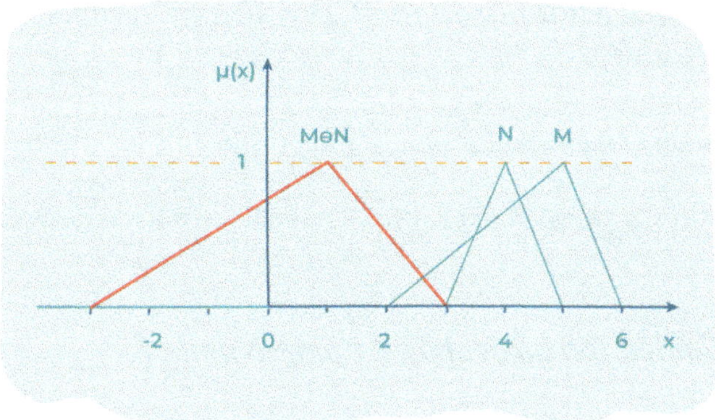

Fig. 3.9 Subtraction $\mathbf{M} \ominus \mathbf{N}$ (bold) of two fuzzy numbers $\mathbf{M}$ and $\mathbf{N}$

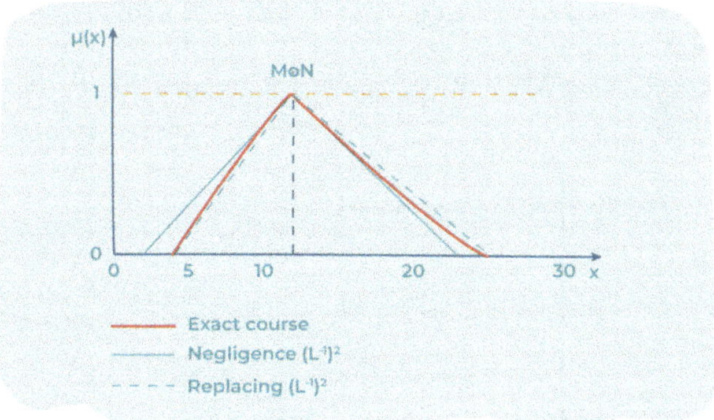

Fig. 3.10 Exact and approximate membership curves for extended multiplication

3.3.5 *Extended Multiplication of Two Fuzzy Numbers* $(\mathbf{M} > 0 \wedge \mathbf{N} < 0) \vee (\mathbf{M} < 0 \wedge \mathbf{N} > 0)$

In order to perform the multiplication $\mathbf{M} \odot \mathbf{N}$ using simple formulas, both factors must be of the *same LR type*. We assume that $\mathbf{M} = \langle m; \alpha; \beta \rangle_{LR} > 0$ and $\mathbf{N} = \langle n; \gamma; \delta \rangle_{RL} < 0$. Using the identity

$$\mathbf{M} \odot \mathbf{N} = \ominus\big[\mathbf{M} \odot (\ominus\mathbf{N})\big], \tag{3.12a}$$

analogous to (3.9) and (3.11), we obtain the approximation

$$\langle m; \alpha; \beta \rangle_{LR} \odot \langle n; \gamma; \delta \rangle_{RL} = \langle mn;\ m\gamma - n\beta + \beta\gamma;\ m\delta - n\alpha - \alpha\delta \rangle_{RL}. \tag{3.12b}$$

3.3.6   Extended Multiplication of Two Fuzzy Numbers $M, N < 0$

If both factors are of the same LR type, the approximation is given by (3.12b):

$$\langle m; \alpha; \beta \rangle_{LR} \odot \langle n; \gamma; \delta \rangle_{LR} = \langle mn; \; -m\delta - n\beta - \beta\delta; \; -m\gamma - n\alpha - \alpha\gamma \rangle_{RL}. \qquad (3.13)$$

3.3.7   Extended Reciprocal of a Fuzzy Number

The reciprocal value $\mathbf{M}^{-1}$ of a fuzzy number is calculated according to the extension principle or according to Theorem 3.3 as follows:

$$\mu_{M^{-1}}(x) = \mu_M(1/x) \qquad \forall x \in \mathbb{R} \setminus \{0\}. \qquad (3.14)$$

For the reciprocal value $\mathbf{M}^{-1}$ of a positive LR number $\mathbf{M} = \langle m; \alpha; \beta \rangle_{LR}$, the following applies:

$$\mu_{M^{-1}}(x) = \begin{cases} R[(1/x - m)/\beta] & x \leq 1/m, \\ & \text{for} \\ L[(m - 1/x)/\alpha] & x \geq 1/m, \end{cases} \qquad (3.15a)$$

$$= \begin{cases} R[(1 - mx)/(\beta x)] & x \leq 1/m, \\ & \text{for} \\ L[(mx - 1)/(\alpha x)] & x \geq 1/m, \end{cases} \qquad (3.15b)$$

Accordingly, $\mathbf{M}^{-1}$ is neither an LR number nor an RL number of the required old type. Similar to multiplication, however, approximation formulas can also be given here.

Approximation for the Vicinity of the New Mean Value $x = 1/m$

The following applies:

$$(x - 1/m)/(\alpha x/m) \approx (x - 1/m)/(\alpha/m^2),$$

$$(m - 1/x)/(\beta x/m) \approx (m - 1/x)/(\beta/m^2),$$

Therefore:

$$\mathbf{M}^{-1} = \langle m; \alpha; \beta \rangle_{LR^{-1}} \approx \langle 1/m; \beta/m^2; \alpha/m^2 \rangle_{RL}. \qquad (3.16)$$

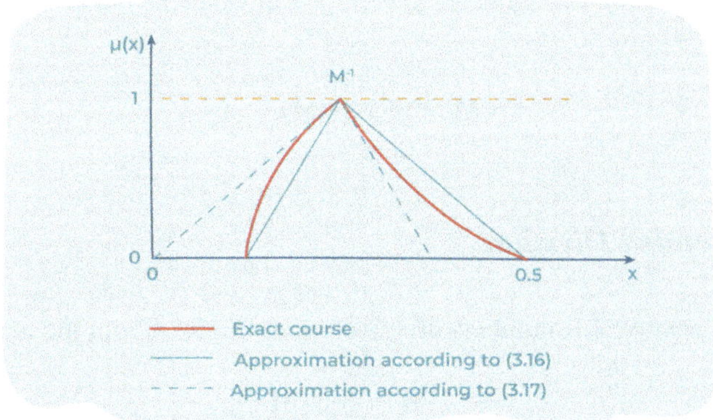

Fig. 3.11 Exact and approximate membership functions for the reciprocals

For the case $x > 1/m$, this approximation is always smaller than the exact solution, and for the case $x < 1/m$, it is always larger. It is therefore left-skewed. This is also expressed in the calculation of the support sets of M^{-1}:

$$\text{Exact solution:} \qquad S(\mathbf{M}^{-1}) = \{x \mid 1/(m + \beta) < x < 1/(m - \alpha)\},$$

$$\text{Approximation (3.16):} \qquad S(\mathbf{M}^{-1}) = \{x \mid 1/m - \beta/m^2 < x < 1/m + \alpha/m^2\}.$$

The above statement is proven with $1/(m + \beta) > 1/m - \beta/m^2$ and $1/(m - \alpha) > 1/m + \alpha/m^2$.

For α and β that are large (compared to m), an approximation can be given, as with multiplication, which is at least at the three points:

$$P_1 = (1/m; 1), \quad P_2 = \big(1/(m + \beta); \ R(1)\big), \quad P_3 = \big(1/(m - \alpha); \ L(1)\big),$$

corresponds to the exact solution. According to Dubois and Prade (1980), it is:

$$\mathbf{M}^{-1} \approx \langle 1/m; \ \beta/m^2 \left[1 - \beta/(m + \beta)\right]; \ \alpha/m^2 \left[1 + \alpha/(m - \alpha)\right]\rangle_{\text{RL}}. \qquad (3.17)$$

Example 3.14 Let $\mathbf{M} = \langle 4; 2; 4\rangle_{\text{LR}}$ with $L(u) = R(u) = \max[0, 1 - u]$. Then the following approximate solutions for $\mathbf{M}^{-1}$ are obtained (Fig. 3.11):

Approximation according to (3.16) : $\mathbf{M}^{-1} \approx \langle 0.25; \ 0.25; \ 0.125\rangle_{\text{RL}},$

$$S(\mathbf{M}^{-1}) = \{x \mid 0 \le x \le 0.375\}.$$

Approximation according to (3.17) : $\mathbf{M}'^{-1} \approx \langle 0.25; \ 0.125; \ 0.25\rangle_{\text{RL}},$

$$S(\mathbf{M}'^{-1}) = \{x \in \mathbb{R} \mid 0.125 \le x \le 0.5\}.$$

Similar approximation formulas for the reciprocal value can be obtained for $\mathbf{M} < 0$ with the identity:

$$\mathbf{M}^{-1} = -(-\mathbf{M})^{-1}.$$

3.3.8   Extended Division

Formulas for positive LR numbers are given as examples. From the identity:

$$\mathbf{N} \oslash \mathbf{M} = \mathbf{N} \odot \mathbf{M}^{-1},$$

it follows (3.10) and (3.16)

$$\langle n; \gamma; \delta \rangle_{\mathrm{LR}} \oslash \langle m; \alpha; \beta \rangle_{\mathrm{LR}} \approx \langle n/m; \ (n\beta+m\gamma)/m^2; \ (n\alpha+m\delta)/m^2 \rangle_{\mathrm{LR}}. \qquad (3.18)$$

and with (3.11) and (3.17)

$$\langle n; \gamma; \delta \rangle_{\mathrm{LR}} \ \odot \ \langle m; \alpha; \beta \rangle_{\mathrm{RL}} \ \approx$$

$$\approx \ \langle n/m; \ (n\beta + m\gamma)(1 - \beta/(m + \beta))/m^2; \ (n\alpha + m\delta)(1 + \alpha/(m - \alpha))/m^2 \rangle_{\mathrm{LR}}$$
$$(3.19)$$

Example 3.15 Let $N = \langle 8; 2; 2 \rangle_{\mathrm{LR}}$ and $M = \langle 4; 1; 2 \rangle_{\mathrm{LR}}$ be given with the shape functions $L(u) = R(u) = \max[0, 1 - u]$. The following approximations can be calculated:

Approximation according to (3.18): $\qquad N \oslash M \approx \langle 2; 1.5; 1 \rangle_{\mathrm{LR}}.$

Approximation according to (3.19): $\qquad (N \oslash M)' \approx \langle 2; 1; 1.33 \ldots \rangle_{\mathrm{LR}}.$

The curve of the membership functions is shown in Fig. 3.12.

In order to be able to approximate membership functions, which are specified by measurement series, for example, more accurately using simple means, fuzzy intervals are often used instead of fuzzy numbers. For this reason, an LR representation is also provided here.

Definition 3.9 (LR Interval) A fuzzy interval $\mathbf{M}$ is called an LR interval if its membership function can be represented by:

$$\mu_M(x) = \begin{cases} L[(m_1 - x)/\alpha] & x \le m_1, \\ 1 & \text{for} \quad m_1 < x \le m_2, \\ R[(x - m_2)/\beta] & m_2 < x, \end{cases}$$

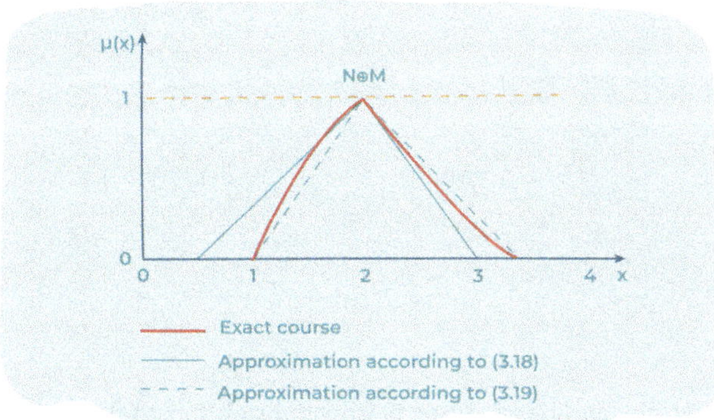

Fig. 3.12 Exact and approximate membership functions for the reciprocals

where $L(u)$ and $R(u)$ are suitable shape functions according to Definition 3.8. The width $m_2 - m_1$ is referred to as the tolerance range of the LR interval.

The abbreviated notation

$$\mathbf{M} = \langle m_1;\ m_2;\ \alpha;\ \beta \rangle_{LR}$$

is used. However, a second representation is also common, which is written as follows:

$$\mathbf{M}' = \langle m';\ c';\ \alpha;\ \beta \rangle_{LR}$$

with $\qquad m' = (m_1 + m_2)/2, \quad c' = (m_2 - m_1)/2.$

Example 3.16 Let the LR interval be $\mathbf{M} = \langle 2;\ 4;\ 1;\ 2 \rangle_{LR} = [3;\ 1;\ 1;\ 2]_{LR}$ with the two branches of type $L(u) = R(u) = \max[0, 1 - u]$. $\mathbf{M}$ can be represented as in Fig. 3.13.

For example, the following calculation rules apply to positive LR intervals:

$$\langle m_1;\ m_2;\ \alpha;\ \beta \rangle_{LR} \oplus \langle n_1;\ n_2;\ \gamma;\ \delta \rangle_{LR}$$

$$= \langle m_1 + n_1;\ m_2 + n_2;\ \alpha + \gamma;\ \beta + \delta \rangle_{LR}, \tag{3.20}$$

$$\langle m_1;\ m_2;\ \alpha;\ \beta \rangle_{LR} \odot \langle n_1;\ n_2;\ \gamma;\ \delta \rangle_{LR} \approx$$

$$\approx \langle m_1 n_1;\ m_2 n_2;\ m_1\gamma + n_1\alpha - \alpha\gamma;\ m_2\delta + n_2\beta + \beta\delta \rangle_{LR}. \tag{3.21}$$

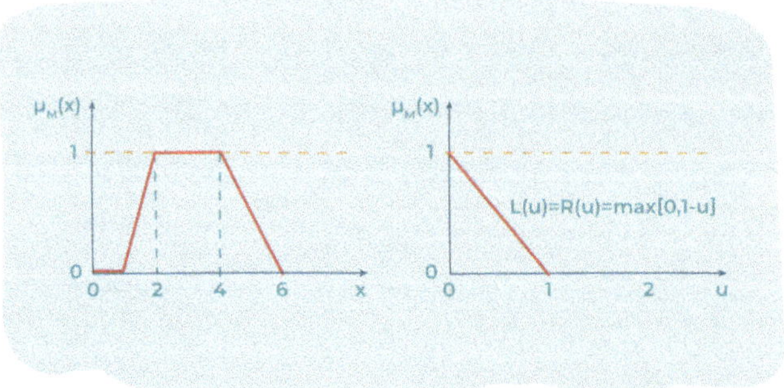

Fig. 3.13 Membership function and shape functions of an LR interval

Example 3.17 Let the two LR intervals be $\mathbf{M} = \langle 4;\ \ 8;\ \ 2;\ \ 4 \rangle_{\mathrm{LR}}$ and $\mathbf{N} = \langle 10;\ \ 14;\ \ 4;\ \ 6 \rangle_{\mathrm{LR}}$. Then, according to (3.20) and (3.21), the following applies:

$$\mathbf{M} \oplus \mathbf{N} = \langle 14;\ \ 22;\ \ 6;\ \ 10 \rangle_{\mathrm{LR}},$$

$$\mathbf{M} \odot \mathbf{N} = \langle 40;\ \ 112;\ \ 28;\ \ 128 \rangle_{\mathrm{LR}}.$$

In particular not only in multiplication but also in addition, considerable fuzziness can quickly arise according to (3.20) and (3.21) by expanding the tolerance ranges of the factors. For further properties of fuzzy numbers and extended operations such as differentiation and integration in fuzzy number spaces, reference is made to the monographs by Dubois and Prade (1980), Klir and Folger (1988), and Zimmermann (1991).

References

Bothe, H.-H. (1995). *Bewertung mit unscharfen Mengen*. Technische Hochschule Ilmenau.

Dubois, D., & Prade, H. (1979). Fuzzy real numbers and fuzzy numerical quantities. *Fuzzy Sets and Systems, 2*(4), 307–318. https://doi.org/10.1016/0165-0114(79)90021-9.

Dubois, D., & Prade, H. (1980). Fuzzy sets and systems: A new approach to the analysis of uncertainty. *Fuzzy Sets and Systems, 3*(1), 3–33. https://doi.org/10.1016/0165-0114(80)90006-0.

Klir, G. J., & Folger, T. A. (1988). *Fuzzy sets, uncertainty, and information*. Prentice Hall.

Rommelfanger, H. (1988). *Fuzzy decision support systems*. Springer.

Zadeh, L. A. (1965). Fuzzy sets. *Information and Control, 8*, 338–353. https://doi.org/10.1016/S0019-9958(65)90241-X.

Zadeh, L. A. (1973). Outline of a new approach to the analysis of complex systems and decision processes. *IEEE Transactions on Systems, Man, and Cybernetics, SMC-3*(1), 28–44. https://doi.org/10.1109/TSMC.1973.5408575.

Zimmermann, H.-J. (1991). *Fuzzy set theory – and its applications* (2nd ed.). Kluwer Academic Publishers.

Chapter 4
Linguistic Expressions

Abstract This chapter introduces linguistic expressions. These are natural language terms to describe variables. Linguistic expressions represent imprecise, subjective, and human-like concepts to allow machines to process vague information.

Communication between experts and users with a rule-based fuzzy expert system can take place in quasi-natural language based on linguistic expressions. Fuzzy controllers and classifiers represent particularly simple rule-based fuzzy expert systems (e.g., see Bothe, 1995). These are described in Chaps. 8 and 9. Linguistic expressions consist of linguistic variables that are linked to each other by linguistic operators. The values of a linguistic variable are mapped to a corresponding numerical value scale by fuzzy sets and thus quantified; the linguistic operators are described by the linking operators for fuzzy sets. This transformation will be explained in more detail below.

4.1 Linguistic Variables

First, the term "linguistic variable" is defined beyond the brief description in Sect. 1.2.

Definition 4.1 (Linguistic Variable (LV)) A linguistic variable is a set system $V_L = \{A, X, G, B\}$ with:

- A set G of syntactic rules that (e.g., in the form of a grammar) determine the linguistic discretization of V_L and thus define the number and nature of the linguistic values of V_L (the terms $\alpha_i \ i \in \mathbb{N}$),
- A set A of terms α_i resulting from G
- A (physically relevant) base set X with the numerical elements $x \in X$
- A set B of semantic rules that assign each term its (physical) meaning in the form of a fuzzy set $\mathbf{M}_{\alpha_i}$ over the base set X

© The Author(s), under exclusive license to Springer Nature Switzerland AG 2026

H.-H. Bothe, E. Portmann, *Computing with Words*, Fuzzy Management Methods,

https://doi.org/10.1007/978-3-032-24117-7_4

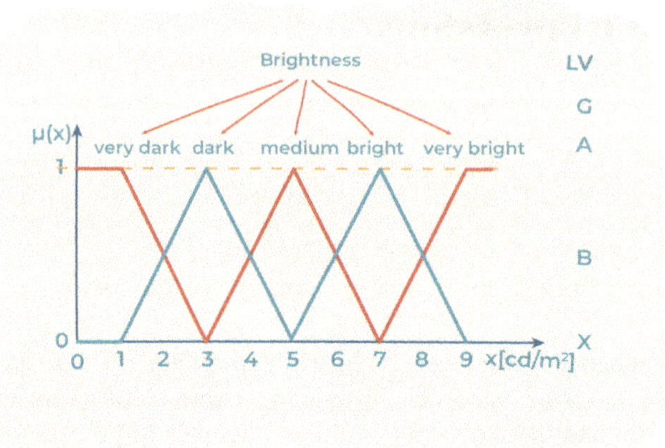

Fig. 4.1 Representation of the terms of the LV "brightness" over the numerical value scale in [cd/m^2]

Example 4.1 Figure 4.1 shows a possible description of the LV "brightness."

According to Fig. 4.1, the LV "brightness" initially includes the set A of possible terms α_i, with:

$$A = \{\text{very dark, dark, medium, bright, very bright}\}.$$

The set G represents the rules according to which the terms α_i are generated (i.e., the terms *dark*, *medium*, and *bright* are specified, and the two outer terms are varied with the word *very*).

Set B describes the transformation of the linguistic value scale created by applying G into a mathematically manageable numerical value scale with physical meaning. Each term $\alpha \in A$ is described by a fuzzy set $\mathbf{M}_\alpha$ over the base set $X = \{x \in \mathbb{R} \mid 0 \le x \le 10\}$ of the relevant numerical brightness values x (in [cd/m^2]). $\mathbf{M}_\alpha$ generates the physical meaning of the term α. The membership values $\mu_\alpha(x)$ can be regarded as the degree of truth that a numerical value x belongs to α.

As an example, the transformation of the term *bright* (shown in Fig. 4.1) is described by the fuzzy set:

$$\mathbf{M}_{\text{bright}} = \{(x; \mu_{\text{bright}}(x)) \mid x \in [0, 9],\ x \in \mathbb{R}\} \quad \text{with}$$

$$\mu_{\text{bright}}(x) = \begin{cases} 0 & x \le 5 \vee x \ge 9, \\ \tfrac{1}{2}(x - 5) & \text{for} \quad 5 < x < 7, \\ \tfrac{1}{2}(9 - x) & 7 < x < 9. \end{cases}$$

The statement "$x = 6$ cd/m^2 belongs to the term *bright*" has a truth value $\mu_{\text{bright}}(6) = \frac{1}{2}$, or in short: The membership value of $x = 6$ cd/m^2 to the term *bright* is $\frac{1}{2}$.

Example 4.2 Linguistic value discretization can be clearly illustrated by the "colors of the rainbow," which are objectively based on a continuous wavelength spectrum in the visible range of light. Subjective color perception extracts colors such as *red*, *orange*, *yellow*, *green*, *blue*, and *violet*. Only this quantization enables the definition and thus the perception of a "color." It structures the concept, which is continuous according to physical standards, through further phenomena such as "'color mixing," "color transition," etc. People from different cultural backgrounds differ in their perception of both the number of colors in a rainbow and the corresponding wavelength range. In most cases, the nature of continuous color transitions does not allow for sharp boundaries between areas, but rather requires blurred transition areas.

The meaning of the quantities G, A, B, and X can also be illustrated using this example.

4.2 Linguistic Operators

A structured form of LV results when the terms and their meanings are determined algorithmically. Linguistic modification operators (in short: modifiers) of the form *very*, *quite*, *quite a lot*, *very very*, etc. are then used.

Definition 4.2 (Linguistic Modification Operator) A linguistic modification operator is an operator on a linguistic value scale. Applying it to a term α results in a new term α'.

If the term is represented by a fuzzy set $\mathbf{A} = \{(x; \mu_A(x))\}$, then the transformation described in Definition 4.2 can be described on the numerical value scale using the set operator concentration CON($\mathbf{A}$), dilation DIL($\mathbf{A}$), complement formation $(\mathbf{A})^c$, and contrast intensification INT($\mathbf{A}$). The membership function $\mu_A(x)$ changes for all $x \in X$ as follows:

$$\textbf{Concentration}: \quad \mu_{\text{CON}}(x) = [\mu_A(x)]^2. \tag{4.1}$$

$$\textbf{Dilation}: \quad \mu_{\text{DIL}}(x) = [\mu_A(x)]^{\frac{1}{2}}. \tag{4.2}$$

$$\textbf{Complementformation}: \quad \mu_{\text{AC}}(x) = 1 - \mu_A(x). \tag{4.3}$$

$$\textbf{Contrastintensification}: \quad \mu_{\text{INT}}(x) = \begin{cases} 2[\mu_A(x)]^2 & \text{for } \mu_A(x) \in [0, 0.5], \\ 1 - 2[1 - \mu_A(x)]^2 & \text{otherwise} \end{cases} \tag{4.4}$$

The application of a modifier to a term α (i.e., the modification of α) can be done, for example, according to the following rules (they are applicable if the position of the mean values or tolerance ranges is not to be shifted):

$$\text{very } \alpha \;\rightarrow\; \text{CON}(\mathbf{A}), \tag{4.5}$$

$$\text{very very } \alpha \;\rightarrow\; \text{CON}[\text{CON}(\mathbf{A})], \tag{4.6}$$

$$\text{quite } \alpha \;\rightarrow\; \text{DIL}(\mathbf{A}), \tag{4.7}$$

$$\text{more than } \alpha \;\rightarrow\; \mathbf{MA} = \{x;\ \mu_{\text{MA}}(x)\ |\ \mu_{\text{MA}}(x) = \mu_A(x)^{1.25}\}, \tag{4.8}$$

$$\text{quite a lot } \alpha \;\rightarrow\; \text{INT}\big[\mathbf{MA} \cap \text{CON}(\mathbf{A})^c\big], \tag{4.9}$$

$$\text{not } \alpha \;\rightarrow\; (\mathbf{A})^c. \tag{4.10}$$

Example 4.3 The definition of LV "truth" (according to Baldwin, 1979) is based on a set A_{truth} of possible terms with:

$$A_{\text{truth}} = \{\text{absolutely false, very false, false, somewhat false, undecided,}$$

$$\text{somewhat true, true, very true, absolutely true}\}.$$

The base set $X = \{v \in \mathbb{R} \mid 0 \leq v \leq 1\}$ is formed by the numerical truth values v, which can take all values between 0 (absolutely false) and 1 (absolutely true) in the sense of a Łukasiewicz logic with ∞ values (Fig. 4.2).

The membership functions of the terms on X are calculated as:

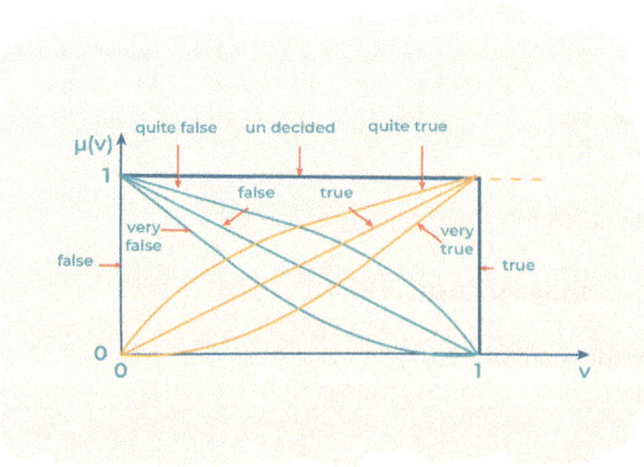

Fig. 4.2 Membership curves for the terms of the LV "truth" according to Baldwin (1979)

$$\mu_{\text{absolutely true}}(v) = \begin{cases} 1 & v = 1, \\ & \text{for} \\ 0 & \text{otherwise}, \end{cases}$$

$$\mu_{\text{absolutely false}}(v) = \begin{cases} 1 & v = 0, \\ & \text{for} \\ 0 & \text{otherwise}, \end{cases}$$

$$\mu_{\text{true}}(v) = v,$$

$$\mu_{\text{false}}(v) = 1 - v,$$

$$\mu_{\text{very true}}(v) = v^2,$$

$$\mu_{\text{very false}}(v) = (1 - v)^2,$$

$$\mu_{\text{somewhat true}}(v) = v^{1/2},$$

$$\mu_{\text{somewhat false}}(v) = (1 - v)^{1/2},$$

$$\mu_{\text{undecided}}(v) = 1.$$

If the terms of an LV represent Boolean expressions of the primary terms (e.g., *true*, *false*) and their modifiers, then we speak of Boolean linguistic variables. The linguistic connectives *and* and *or* are used.

Definition 4.3 (Linguistic Linking Operators) The linguistic linking operators *and/or* are binary operators on a linguistic value scale and form the verbal equivalents of the logical operators ($\wedge$) and ($\vee$) of Boolean algebra. They link two linguistic terms α and β of a linguistic value scale. The result is a new linguistic expression γ.

The linguistic linking operators are mapped to the numerical world using the operators for fuzzy sets. If the terms α and β to be linked are represented by the fuzzy sets **A** and **B** via the corresponding basic numerical sets X_1 and X_2, linking α and β produces a fuzzy set **C** via $Y = X_1 \times X_2$, which represents the link result γ.

Example 4.4 Let $A_{\text{age}} = \{\text{young, old, young or old}, \ldots\}$. For the assignment of linguistic operators to operators for fuzzy sets, the following applies:

$$\text{not} \rightarrow \text{complement formation},$$

$$\text{and} \rightarrow \text{intersection},$$

$$\text{or} \ \rightarrow \text{union}.$$

Furthermore, let the terms *young* and *old* be given by:

$$
\mu_{\text{young}}(x) = \begin{cases} 1 & x \in [0, 25], \\[2ex] & \text{for} \\[2ex] \left[1 + \tfrac{1}{4}(x - 25)^2\right]^{-1} & x \in [25, 100], \end{cases}
$$

$$
\mu_{\text{old}}(x) = \begin{cases} 0 & x \in [0, 50], \\[2ex] & \text{for} \\[2ex] \left[1 + \tfrac{1}{4}(x - 50)^2\right]^{-1} & x \in [50, 100]. \end{cases}
$$

Then, for example, the Boolean linguistic expressions *not very old* and *young or old* can be expressed as follows:

$$
\mu_{\text{not very old}}(x) = \begin{cases} 1 & x \in [0, 50], \\[2ex] & \text{for} \\[2ex] 1 - \left[1 + \tfrac{1}{4}(x - 50)^2\right]^{-2} & x \in [50, 100]. \end{cases}
$$

$$
\mu_{\text{young or old}}(x) = \begin{cases} 1 & x \in [0, 25], \\[1ex] \left[1 + \tfrac{1}{4}(x - 25)^2\right]^{-1} & \text{for } x \in [25, c], \\[1ex] \left[1 + \tfrac{1}{4}(x - 50)^2\right]^{-1} & x \in [c, 100], \end{cases}
$$

with $c = \tfrac{1}{2}(75 + 626^{1/2})$.

Comments

1. When transforming linguistic operators, attention must be paid to both the order in which they are applied and their nesting. For example, the expression not very (young or old) has a different membership function than not ((very young) or old). To transform such a compound expression, its exact meaning must therefore be known.
2. Language can sometimes be illogical and therefore disrupt the system of a linguistically defined scale. For instance, a young man is young, but in everyday language, a "younger" man is older than a young man if "younger" and "young" are treated as absolute values.
3. Not only must the terms themselves be well defined but also the linguistic linking operators.

References

Baldwin, J. F. (1979). A new approach to approximate reasoning using fuzzy logic. *Fuzzy Sets and Systems*, 2(4), 309–325. https://doi.org/10.1016/0165-0114(79)90022-0
Bothe, H.-H. (1995). *Bewertung mit unscharfen Mengen*. Technische Hochschule Ilmenau.

Chapter 5
Probability and Possibility

Abstract This chapter is an annex of probability to possibility theory. While both allow for dealing with uncertainty, the former focuses on the likelihood of an event occurring, whereas the latter deals with the potential for an event to occur when information is incomplete or vague.

While the measures described in Sect. 2.4 make the specific fuzziness of fuzzy sets comparable with each other, the question here is to what extent a particular value belongs to a (crisp) set. A qualitative (i.e., subjective) statement by an expert can serve as a starting point, which must then be translated into a mathematical formalism. This gives rise to the class of fuzzy measures. The most important fuzzy measures are probability and possibility (e.g., see Bothe, 1995).

5.1 Principle of Incompatibility

Often imprecision cannot be measured directly, and, in addition, there might be situations in which, even if it could be measured, the results would be of little use or relatively difficult to interpret. Hence, an aggregated model sometimes yields more immediately assimilated information than a very detailed, and therefore more exact, model. In this sense, Zadeh (1973) ascertained that "as the complexity of a system increases, our ability to make precise and yet significant statements about its behavior diminishes until a threshold is reached beyond which precision and significance (or relevance) become almost mutually exclusive characteristics."

This principle of incompatibility is linked to the way we humans perceive and reason: We largely use summary representations of reality, which are therefore imprecise and generally tainted with subjectivity. In fact, a "good model" must achieve a compromise that avoids any excessive precision and is likely to be arbitrary and so uncertain. In this sense, also the concept of chance, which was associated with randomness, and the concept of probability, as an attribute of opinion, were both viewed independently of each other. Thus, the principle of

H.-H. Bothe, E. Portmann, *Computing with Words*, Fuzzy Management Methods,
https://doi.org/10.1007/978-3-032-24117-7_5

chance and the calculus of probabilities were developed individually (Dubois & Prade, 1988).

The success of the theory of probability is strongly linked to the expansion of physical science, in which the problems of the modeling of judgment faded into the background. Concerns led to what was marked by a partiality for the additivity of degrees of belief. As a solution, the economist Shackle (1961) proposed a non-probabilistic framework, in which our human reasoning processes are analyzed in terms of "possibility." By assessing plausibilities (i.e., the natural process of creatively balancing possible alternative outcomes through human perception and in human reasoning), Shackle integrated "imagination" into decision-making. Subsequently, his notions of upper and lower probabilities were introduced to deal with incomplete data, but they were no longer additive. In Sect. 5.2, we shall thus contrast probability with possibility.

5.2 Probability Versus Possibility

For mathematical formulation, we start from a basic domain X. Let $Q(X)$ be a system of subsets A_i of X (see Definition 2.3: Power set). For $X = N$, let $Q(X) = \{\{1\}, \{1, 2\}, \{2, 3, 4\}\}$. Then a fuzzy measure can be regarded as a set function $F_Q(A)$ that maps the set system $Q(X)$ onto the unit interval: $F_Q : Q(X) \rightarrow [0, 1]$. It thus corresponds to the membership function $\mu_A : X \rightarrow [0, 1]$ when defining fuzzy sets, which evaluates the individual elements of X with regard to their membership in A.

For ease of mathematical handling, $Q(X)$ is usually replaced by a suitable σ algebra $\mathcal{L}(X)$ over X. This implies the following additional conditions for $\mathcal{L}(X)$:

1. $X \in \mathcal{L}(X)$.
2. $A \in \mathcal{L}(X) \Rightarrow A^c \in \mathcal{L}(X)$.
3. $A_i \in \mathcal{L}(X), i = 1, \ldots, n \Rightarrow \cup_i A_i \in \mathcal{L}(X)$.

When introducing fuzzy measures, it is assumed that the sets to be measured $A \in \mathcal{L}(X)$ represent certain events. Each event A is then assigned a measure value $F(A)$. A σ algebra $\mathcal{L}(X)$ is also referred to as the event space on the result space X.

Example 5.1 Let $X = \{1, 2, 3, 4, 5, 6\}$ be the set of possible outcomes x when rolling a die on which a set system $\mathcal{L}(X) = \{\varnothing, \{1, 3, 5\}, \{2, 4, 6\}, X\}$ with the events "invalid," "odd number," "even number," and "valid" is defined. $\mathcal{L}(X)$ represents a σ–algebra over the base set $X = \{1, 2, 3, 4, 5, 6\}$. The event $A = \{1, 3, 5\}$, which means the "odd number," can be assigned a measure $F(A)$ that provides information about the extent to which a particular roll of the die with the result x leads to event A. For a hypothetical, initially unknown die result x, both a probability value $F_1(A) = 0.5$ and a possibility value $F_2(A) = 1$ can be specified. However, if there is no knowledge of the physical background, an expert could also determine the possibility value $F^*(A) = 0.9$ based on their (supposed) experience.

This subjective decision must be taken into account; it reflects the expert's personal knowledge of the handling of a process (e.g., an asymmetry of the die), which, based on experience, does not necessarily require explicit system knowledge.

Definition 5.1 (Fuzzy Measure) A set function $F(A)$ is called a fuzzy measure on $\mathcal{L}(X)$ if the following applies to all $A_i \in \mathcal{L}(X)$:

1. $F(\emptyset) = 0$ and $F(X) = 1$.   **[Normalization]**
2. $A_1 \subseteq A_2 \Rightarrow F(A_1) \leq F(A_2)$.   **[Monotonicity]**
3. $A_1 \subseteq A_2 \subseteq \cdots \Rightarrow \lim_{n\to\infty} F(A_n) = F(\lim_{n\to\infty} A_n)$.   **[Continuity]**

If X, $\mathcal{L}$, and F are specified, the result of the measurement is uniquely determined. Then $[X, \mathcal{L}, F]$ forms a fuzzy measure space.

Remarks

1. Definition 5.1(2) means that the measure value cannot decrease when the set is extended.
2. For finite basic domains, Definition 5.1(3) is not strictly necessary in order to construct a meaningful theory of fuzzy measures.

From Definition 5.1, the following immediately follows for two sets $A_1, A_2 \in \mathcal{L}(X)$:

Theorem 5.1

1. $F(A_1 \cup A_2) \geq \max[F(A_1).F(A_2)]$,
2. $F(A_1 \cap A_2) \leq \min[F(A_1), F(A_2)]$.

Definition 5.2 (Probability) A function $P : \mathcal{L}(X) \to [0, 1]$ is called probability on $\mathcal{L}(X)$ if the following applies to all $A_i \in \mathcal{L}(X)$:

1. $P(A_i) \geq 0$.
2. $P(X) = 1$.
3. $\forall i \neq j : A_i \cap A_j = \emptyset \Rightarrow P(\cup_i A_i) = \sum_i P(A_i)$.

For finite base sets X, condition 3 (i.e., the addition principle) can be weakened to:

3′. $A_1, A_2 \in \mathcal{L}(X), \ A_1 \cap A_2 = \emptyset \Rightarrow P(A_1 \cup A_2) = P(A_1) + P(A_2)$.

The probability defined in this way represents a special fuzzy measure. The restriction is that the monotonicity requirement for general fuzzy measures has been replaced by the stronger additivity requirement.

According to Definition 5.2, each event A from $\mathcal{L}(X)$ is assigned a probability $P(A)$, which for finite or countably infinite base sets (e.g., $\mathbb{N}$) can be written as the sum of the probabilities $P(\{x\})$ of the individual events $\{x\}$ as:

$$P(A) = \sum_{x \in A} P(\{x\}). \tag{5.1}$$

Using the characteristic function $\mu_A(x)$ of A on the base set X, (5.1) is transformed to:

$$P(A) = \sum_{x \in X} \mu_A(x)\, P(\{x\}). \tag{5.2}$$

In integral notation, $P(A)$ can be expressed as:

$$P(A) = \int_X \mu_A(x)\, dP(x) = \int_X \mu_A(x)\, p(x)\, dx, \tag{5.3}$$

where $p(x)$ is the probability density distribution.

Zadeh (1968) generalized (5.2) to the probability of fuzzy events by also interpreting $\mu_A(x)$ as the membership function of a fuzzy set $\mathbf{A}$. This means introducing probability for the terms α of a linguistic variable. Because

$$\sum_{x \in X} P(\{x\}) = P(X) = 1 \text{ and} \tag{5.5}$$

$$\int_X p(x)\, dx = 1, \tag{5.6}$$

Eqs. (5.2) and (5.3) describe a probability-weighted averaging of the membership function $\mu_A(x)$ over the basic domain X. The probability $P(\mathbf{A})$ of a fuzzy set $\mathbf{A}$ can therefore be interpreted as the average membership value of the elements $x \in X$ to $\mathbf{A}$.

The approach described can be extended by averaging several membership functions (i.e., designed by different experts) μ_A for the same term α to establish "probable function curves." For this purpose, Hirota (1977) introduced probability sets A in which the membership functions $\mu_A(x, \omega)$ for a term depend both on the elements x of the basic domain X and on the elements ω of a sample space Ω (i.e., the experts ω). Due to (5.5), an expectation function can be easily obtained by averaging

$$E(\mu_A)(x) = \int_\Omega \mu_A(x, \omega)\, dP(\omega) \quad \forall x \in X \tag{5.7}$$

and a variance function

$$V(\mu_A)(x) = \int_\Omega [\mu_A(x, \omega) - E(\mu_A)(x)]^2\, dP(\omega) \quad \forall x \in X \tag{5.8}$$

for A can be specified by averaging. In this way, membership functions with a measurable "certainty" are created.

Definition 5.3 (Possibility) A function $\Pi : \mathcal{L}(X) \to [0, 1]$ is called a possibility on $\mathcal{L}(X)$ if the following applies for all $A_i \in \mathcal{L}(X)$, $i = 1, \ldots, n$:

1. $\Pi(\varnothing) = 0$.
2. $\Pi(X) = 1$.
3. $\Pi(\cup_i A_i) = \sup_i \Pi(A_i)$.

For finite sets X, condition 3 can be weakened to:

3′. $A_1, A_2 \in \mathcal{L}(X)$, $A_1 \cap A_2 = \varnothing \Rightarrow \Pi(A_1 \cup A_2) = \max[\Pi(A_1), \Pi(A_2)]$.

The possibility defined in this way is a special fuzzy measure if and only if X is finite. Here, the monotonicity requirement for general fuzzy measures is replaced by the stronger supremum requirement.

According to Definition 5.3, each event $A \in \mathcal{L}(X)$ is assigned a possibility $\Pi(A)$, which can be expressed using the characteristic function $\mu_A(x)$ of $A \subseteq X$ as follows:

$$\Pi(A) = \sup_{x \in A} \Pi(\{x\}) = \sup_{x \in X} [\mu_A(x)\, \Pi(\{x\})]. \tag{5.9}$$

From (5.9) and Definition 5.3(2), it immediately follows that:

$$\sup_{x \in X} \Pi(\{x\}) = 1. \tag{5.10}$$

Definition 5.3 (3′) assigns the smallest possible value to the first part of Theorem 5.1. Thus, the statement "the event is possible" is the weakest formulation of the realization of the event in the sense of fuzzy measure theory.[1]

Theorem 5.2 *The probability values $P(A)$ on the base set X can be regarded as the lower bound for the possibility values $\Pi(A)$, which means:*

$$\Pi(A) \geq P(A) \quad \forall A \in \mathcal{L}(X).$$

According to (5.9), the following must therefore apply:

$$\Pi(\{x\}) \geq P(\{x\}) \qquad \forall x \in X.$$

Example 5.2 The crisp statement A "Karl eats x rolls for breakfast" is to be assessed in terms of possibility and probability. The base set is assumed to be the set $X = \{0, 1, 2, 3, 4, 5\}$ with the numerical results $x \in X$. The individual events $A_0 = \{0\}$, $A_1 = \{1\}$, $A_2 = \{2\}$, ..., $A_5 = \{5\}$ are to be measured. How are the measures derived?

Possibility: Through subjective assessment of Karl's behavior and the current process situation (i.e., refrigerator empty).

Probability: By statistical evaluation of the event over a longer period of time with the process situation remaining unchanged. A forecast can lead to the following table:

---

[1] This is a mathematical description of a *subjective* assessment of the possibility of an event occurring, not an *objective* physical possibility.

x	0	1	2	3	4	5
$\Pi(\{x\})$	1	1	1	0.6	0.3	0.1
$P(\{x\})$	0.1	0.2	0.5	0.1	0.1	0

The objectivity of $P(\{x\})$ compared to $\Pi(\{x\})$ can be easily questioned, as it is almost impossible to guarantee a consistent process situation during data collection. For a prediction without precise knowledge of the process situation, the probability distribution $\pi(x) = \Pi(\{x\})$ should therefore be used in the sense of a subjectively estimated worst-case treatment, which can also be understood as the normalized membership function of a fuzzy set $A = \{(x; \mu_A(x)) \mid x \in A\}$; it indicates the probability of the individual elements x belonging to the set A. Statements about probability, on the other hand, would require a complete overview of the system and the associated experience.

Example 5.3 In Example 5.2, the question could arise as to how likely it is that Karl will eat three, four, or five rolls. Using Definition 5.3 (3) and $A = \cup_{x \in A}\{x\}$:

$$\Pi(A) = \Pi(\cup_{x \in A}\{x\}) = \sup_{x \in A} \pi(x).$$

$$A = \{3, 4, 5\} \Rightarrow \Pi(A) = 0.6$$

The possibility $\Pi(A)$ for the occurrence of event A is evaluated as 0.6. The probability $P(A)$, on the other hand, is given by Definition 5.2 (3) as:

$$P(A) = \sum_{x=3,4,5} P(\{x\}) = 0.2 < \Pi(A).$$

In summary, the following applies to $\Pi(\{x\})$ and $P(\{x\})$:

$$\sum_{x \in X} P(\{x\}) = 1 \quad \text{and} \quad \sup_{x \in X} \Pi(\{x\}) = 1,$$

but obviously not necessarily $\sum_{x \in X} \Pi(\{x\}) = 1$.

In addition to probability and possibility, other fuzzy measures can also be defined, or their curve can be given in the form of distribution functions, for example:

- The "guaranteed possibility" as a measure of how much all x belong to A:

$$\Delta(A) = \inf_{x \in A} \pi(x). \tag{5.11}$$

- The "necessity" as a measure of how much each value outside A does not belong to A:

$$N(A) = 1 - \Pi(A^c) = \inf_{x \in A} [1 - \pi(x)]. \tag{5.12}$$

– The "potential certainty" as a measure of the extent to which every value outside A is not guaranteed to belong to A, which means the extent to which at least one value in the complement of A exists with a low possibility value:

$$\nabla(A) = 1 - \Delta(A^c) = 1 - \sup_{x \in A} \pi(x). \qquad (5.13)$$

While $\Delta(A)$ takes into account the possibilities of all values in A according to (5.11), $N(A)$ assesses the certainty of $x \in A$ regarding the impossibility of all values outside A according to (5.12). $\nabla(A)$ is calculated according to (5.13) from the absence of a guaranteed possibility for A^c. Both $\Pi(A)$, $\Delta(A)$, $N(A)$, and $\nabla(A)$ use only the qualitative operators "sup," "inf," and "$1 - (\cdot)$." This indicates that the ranking of the possibility values is of particular importance. According to Dubois and Prade (1992), the following inequality applies to normalized $1 - \pi(x)$ and $\pi(x)$:

$$\max[N(A),\ \Delta(A)] \ \leq \ \min[\Pi(A),\ \nabla(A)]. \qquad (5.14)$$

By interpreting the possible events A, the specified measures can be applied, for example, to evaluate:

– The results of a quality check
– The course of a patient's illness
– The necessity of manual intervention in a process

The possibility measure $\Pi(A)$ and the dual necessity measure $N(A)$ are most commonly used to model measurement results that have been obtained without a guarantee of a consistent process environment or subjective assessments of process states.

References

Bothe, H.-H. (1995). *Bewertung mit unscharfen Mengen*. Technische Hochschule Ilmenau.

Dubois, D., & Prade, H. (1988). *Possibility theory: An approach to computerized processing of uncertainty*. Plenum Press.

Dubois, D., & Prade, H. (1992). Putting fuzziness and uncertainty together. In L. A. Zadeh & J. Kacprzyk (Eds.), *Fuzzy logic for the management of uncertainty* (pp. 3–20). John Wiley & Sons.

Hirota, K. (1977). Concepts of probabilistic fuzzy systems. *Information Sciences, 13*(1), 1–20. https://doi.org/10.1016/0020-0255(77)90059-5.

Shackle, G. L. S. (1961). *Decision, order, and time in human affairs*. Cambridge University Press.

Zadeh, L. A. (1968). Probability measures of fuzzy events. *Journal of Mathematical Analysis and Applications, 23*(2), 421–427. https://doi.org/10.1016/0022-247X(68)90078-4.

Zadeh, L. A. (1973). Outline of a new approach to the analysis of complex systems and decision processes. *IEEE Transactions on Systems, Man, and Cybernetics, SMC-3*(1), 28–44. https://doi.org/10.1109/TSMC.1973.5408575.

Chapter 6
Fuzzy Relations

Abstract This chapter presents fuzzy relations as an extension of a classical relation. They allow for degrees of connection between elements, rather than a simple binary association. A fuzzy relation assigns a membership grade to every pair of elements, representing the strength of their association.

The preceding chapters discussed the conversion of linguistic terms and corresponding linking operators into a mathematically comprehensible language. According to Bothe (1995), this allows the behavior of technical systems to be described first qualitatively and then quantitatively using fuzzy sets. The same applies to the input and output variables of the systems. This chapter will lay the foundations for drawing conclusions based on such fuzzy information. This task arises, for example, when a rule-based expert system is required to respond to input values for which no output values are explicitly specified in the rules. The problem is fundamental in nature, as the number of rules is always limited. The special cases of rule-based fuzzy controllers and classifiers are described in more detail in Chapters 7, 8, and 9.

The process of inference can be described as "approximate reasoning." To understand the methodology, the term "fuzzy relation" is first introduced.

6.1 Properties

The classic concept of a relation (e.g., "$\geq$," "$=$") can be directly transferred to the domain of fuzzy sets. Relations are used, for example, in controller design and classification. Binary fuzzy relations are fuzzy sets that either evaluate the (crisp) elements of the Cartesian product $X \times Y$ according to a fuzzy selection rule or compare two variables with fuzzy values (i.e., fuzzy variables) on a basic domain $X \times Y$.

The first case will be dealt with first.

Definition 6.1 (Fuzzy Relation Between Crisp Sets) Let $X_1, \ldots, X_n \subseteq \mathbb{R}$ be the base sets with the Cartesian product $X_1 \times \cdots \times X_n$. Then the mapping

H.-H. Bothe, E. Portmann, *Computing with Words*, Fuzzy Management Methods,
https://doi.org/10.1007/978-3-032-24117-7_6

Fig. 6.1 Excerpt from the
membership function of the
fuzzy relation $R_{\ll}(x_1, x_2)$:
much smaller than

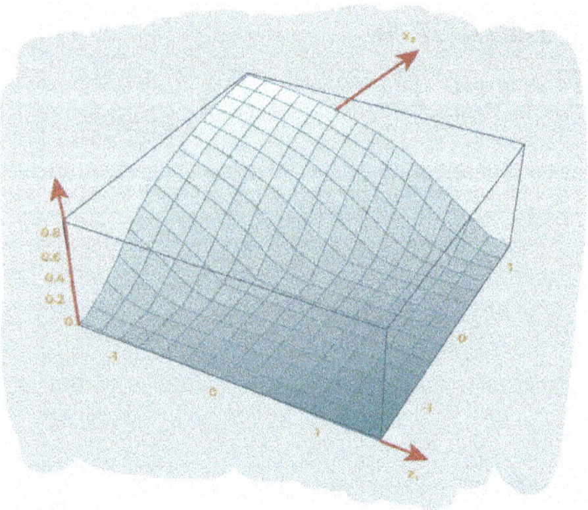

$$R : X_1 \times \cdots \times X_n \to [0, 1]$$

is an n-ary fuzzy relation between the crisp values $x_1, \ldots, x_n$. $R(x_1, \ldots, x_n)$ can
be described by a membership function $\mu_R : (x_1, \ldots, x_n) \mapsto [0, 1]$ and thus be
regarded as a fuzzy set on $X_1 \times \cdots \times X_n$: $\{(x_1, \ldots, x_n); \ \mu_R(x_1, \ldots, x_n)\}$.

Example 6.1 On $X_1 \times X_2 = \mathbb{R}^2$, a binary fuzzy relation $R_{\ll}(x_1, x_2)$: *much smaller
than* can be defined as follows:

$$\mu_{R_{\ll}}(x_1, x_2) = \begin{cases} 0 & \quad x_1 \geq x_2, \\ \left[1 + (x_2 - x_1)^2\right]^{-1} & \text{for} \quad x_1 < x_2. \end{cases}$$

The membership function $\mu_{R_{\ll}}(x_1, x_2)$ for evaluating the value pairs (x_1, x_2)
can be represented in the range $-1.4 < x_1 < 1.4$, $-1.4 < x_2 < 1.4$ (as shown in
Fig. 6.1).

For finite support sets, binary fuzzy relations can also be represented using
membership matrices.

Example 6.2 Representation of the fuzzy relation $R_{\gg}(x_1, x_2)$: *significantly greater
than* ($\gg$) on the Cartesian product $X_1 \times X_2 = \{6, 15, 30\} \times \{1, 2, 5, 10\}$:

$\mu_{R_{\gg}}(x_1, x_2) : x_1$ $\quad$ x_2	1	2	5	10
6	0.5	0.1	0	0
15	1	0.5	0.2	0.1
30	1	0.9	0.5	0.3

Note The membership width of fuzzy relations is determined by subjective assessment, taking into account the physical units in the underlying Cartesian product
space. A basic statement such as the following can be used for this purpose:
"A temperature value T in [K] is *more dangerous* for the functionality of a

semiconductor component than a humidity value F in [mbar]." The membership values for the relation $\mathbf{R}_g$: *more dangerous than* are then determined by assessing the basic statement when varying the value pairs (T, F).

In many cases, it makes sense to define fuzzy relations between fuzzy sets. Linguistic values (or terms, such as *large, fast, hot*) can then also be compared with each other across the product spaces $X_1 \times \cdots \times X_n \subseteq \mathbb{R}^n$. The terms are understood as fuzzy sets. A relation between them is formed on the basis of the Cartesian product for fuzzy sets. Therefore, Definition 6.1 should be expanded in the following.

Definition 6.2 (Fuzzy Relation Between Fuzzy Sets) Let $X_1, \ldots, X_n \subseteq \mathbb{R}$ be the base sets on which the fuzzy sets $\mathbf{A}_i = \{(x_i; \mu_{A_i}(x_i)) \mid x_i \in X_i, \ i = 1, \ldots, n\}$ are defined. Then the mapping

$$R : \mathbf{A_1}, \ldots, \mathbf{A_n} \to [0, 1]$$

is an n-ary fuzzy relation between the fuzzy sets $\mathbf{A}_i$ if $\forall (x_1, \ldots, x_n) \in X_1 \times \cdots \times X_n$ holds: $R \subseteq A_1 \otimes \cdots \otimes A_n$, for example, for $n = 2$:

$$\mu_R(x_1, x_2) \leq \mu_{KP}(x_1, x_2) = \min[\mu_A(x_1), \mu_B(x_2)].$$

$R(A_1, \ldots, A_n)$ again forms a fuzzy set. In the case $n = 2$:

$$R(A_1, A_2) = \{((x_1, x_2) \ ; \ \mu_R(x_1, x_2)) \mid (x_1, x_2) \in X_1 \times X_2\}.$$

Note Both the Cartesian product $\mathbf{A}_1 \otimes \mathbf{A}_2$ of two fuzzy sets $\mathbf{A}_1 \in \mathbf{P}(X_1)$ and $\mathbf{A}_2 \in \mathbf{P}(X_2)$ and the fuzzy relations applicable to them can be represented as a curved surface in a space with x_1, x_2 and $\mu_{KP}(x_1, x_2)$. The condition in Definition 6.2 now requires that all conceivable fuzzy relations on $\mathbf{X}_1 \times \mathbf{X}_2$ are contained in the fuzzy Cartesian product. The corresponding surfaces must therefore have membership values that are less than or equal to those of the Cartesian product $\mathbf{A}_1 \otimes \mathbf{A}_2$. The Cartesian product is the largest possible combination of $\mathbf{A}_1$ and $\mathbf{A}_2$ in terms of inclusion. In the case of two crisp sets $\mathbf{A}_1$ and $\mathbf{B}_2$, which means for $\mu_1(x_1), \mu_2(x_2) \in \{0, 1\}$, the restriction in Definition 6.2 is reduced to the condition $\mu_R(x_1, x_2) \leq 1$.

Example 6.3 Let the terms *fast* and *cold* be given to characterize the linguistic variables "flow velocity" and "temperature" of a coolant on the discretized basic domains $V[\text{m/s}] = \{1, 2, 3, 4, 5\}$ and $T[°C] = \{0, 100, 200, 300, 400, 500\}$. The term *fast* is represented by $V_{\text{fast}} = \{(1; 0), (2; 0.3), (3; 0.9), (4; 1), (5; 1)\}$, and the term *cold* by $T_{\text{cold}} = \{(100; 1), (200; 1), (300; 0.7), (400; 0.2), (500; 0)\}$; the membership functions of the two terms are shown in Fig. 6.2.

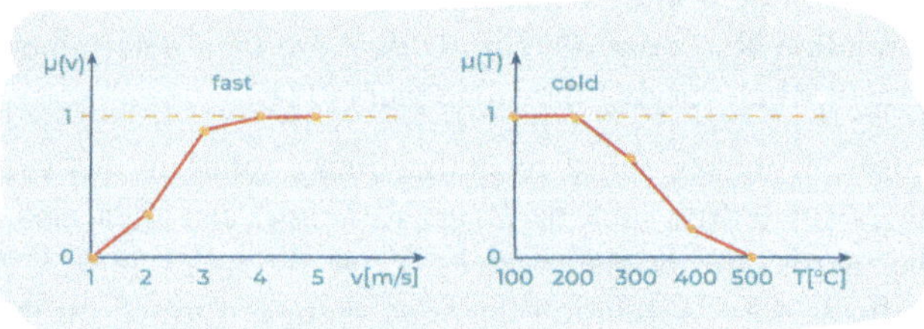

Fig. 6.2 Membership functions of the two terms *fast* and *cold*

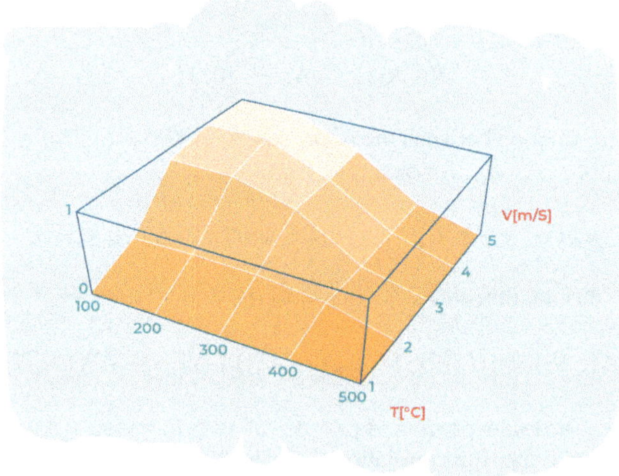

Fig. 6.3 3D representation of the Cartesian product in the velocity-temperature space: *fast-flowing* and at the same time *cold* fluid

The Cartesian product of V_{fast} and T_{cold} then results in:

$\mu_{KP}(v, T):$ $v \backslash T$	100	200	300	400	500
1	0	0	0	0	0
2	0.3	0.3	0.3	0.2	0
3	0.9	0.9	0.7	0.2	0
4	1	1	0.7	0.2	0
5	1	1	0.7	0.2	0

$\mu_{KP}(v, T)$ can be interpreted as the membership function of current state values (v, T) for a *fast-flowing* and simultaneously *cold* coolant. Figure 6.3 shows a 3D representation.

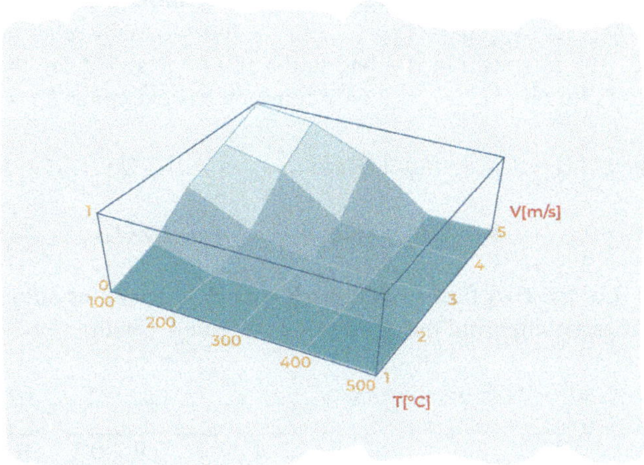

Fig. 6.4 3D representation of the fuzzy relation $R(V_{\text{fast}}, T_{\text{cold}})$: liquid *that flows quickly* rather than being *cold*

On the Cartesian product, an expert can now form, for example, a fuzzy relation $R(V_{\text{fast}}, T_{\text{cold}})$ = "Relatively speaking, the liquid flows *faster* than it is *cold*":

$\mu_R(v, T):$ $v \backslash T$	100	200	300	400	500
1	0	0	0	0	0
2	0.2	0	0	0	0
3	0.6	0.2	0	0	0
4	0.8	0.6	0.1	0	0
5	1	0.9	0.6	0	0.

This fuzzy relation is shown in Fig. 6.4 as a 3D graphic. This allows a direct comparison between the Cartesian product $V_{\text{fast}} \otimes T_{\text{cold}}$ (Fig. 6.2) and the fuzzy relation $R(V_{\text{fast}}, T_{\text{cold}})$.

Fuzzy relations according to Definitions 6.1 and 6.2 represent fuzzy sets, regardless of whether they refer to crisp or fuzzy variables. Their notation can be standardized by always using the names of the corresponding basic domain elements to represent the reference variables. The notation of fuzzy relations $R(x_1, \ldots, x_n)$ between crisp variables $x_1 \in X_1, \ldots, x_n \in X_n$ is then retained, while fuzzy relations $R(u_1, \ldots, u_n)$ between fuzzy variables $u_1, \ldots, u_n$ on the basic domains $X_1, \ldots, X_n$ are also written as $R(x_1, \ldots, x_n)$. In the following, this uniform notation is used, except when explicit reference is to be made to the fuzzy reference variables or when the reference variables of a relation have already been mentioned elsewhere and are not of essential importance; in such cases, their representation is omitted in order to facilitate access through simple notation.

Definition 6.3 (Intersection and Union of Fuzzy Relations in the Same Product Space) Let $R(x_1, \ldots, x_n)$ and $Z(x_1, \ldots, x_n)$ be the two fuzzy relations in the same product space. The intersection ($\cap$) and union ($\cup$) of R and Z are then defined as linking operators for all $(x_1, \ldots, x_n)$ on a common basic domain $X_1 \times \cdots \times X_n$ by:

$$\mu_{R \cap Z}(x_1, \ldots, x_n) = \min\{\mu_R(x_1, \ldots, x_n), \mu_Z(x_1, \ldots, x_n)\},$$

$$\mu_{R \cup Z}(x_1, \ldots, x_n) = \max\{\mu_R(x_1, \ldots, x_n), \mu_Z(x_1, \ldots, x_n)\}.$$

Example 6.4 Let the two fuzzy relations **R** and **Z** be given as subjective assessments by an expert comparing the crisp values of x and y with:

$R(x, y) =$ "x is greater than y" :

$x \backslash y$	1	2	5	10
4	1	0.9	0.1	0
8	1	1	0.6	0.1
15	1	1	1	0.6

$Z(x, y) =$ "x is approximately equal to y" :

$x \backslash y$	1	2	5	10
4	0.1	0.4	0.9	0.1
8	0	0	0.5	0.8
15	0	0	0.1	0.6

The membership values of intersection $R \cap Z$ and union $R \cup Z$ are then calculated as follows:

$\mu_{R \cup Z}(x, y)$:

$x \backslash y$	1	2	5	10
4	1	0.9	0.9	0.1
8	1	1	1	0.8
15	1	1	1	0.6

$\mu_{R \cap Z}(x, y)$:

$x \backslash y$	1	2	5	10
4	0.1	0.4	0.1	0
8	0	0	0.5	0.1
15	0	0	0.1	0.6

The results can be interpreted as follows:

$R \cup Z$: "x is greater than or approximately equal to y,"

$R \cap Z$: "x is greater than and at the same time approximately equal to y."

6.2 Concatenation of Fuzzy Relations

In the next step, fuzzy relations from different product spaces are also to be combined with each other so that, by executing them sequentially, a new fuzzy relation is created on a concatenated product space.

The "composition" or "concatenation" of fuzzy relations represents an extension of the concatenation of crisp relations according to Definition 2.9 and is generally important in decision theory, especially in controller design and automated classification. The fuzziness of the result relation arises from the fuzziness of the relations to be concatenated and from the concatenation rule itself, whereby different calculation rules are conceivable for the quantitative implementation. In contrast to Definition 2.9 and (2.4), however, their results will now differ. The most commonly used is max-min composition, which is a direct extension of (2.4) and is defined below.

Definition 6.4 (Max-Min Concatenation (Max-Min Composition)) The max-min composition ($\circ_{MM}$) of the fuzzy relations

$$R_1(x, y) = \{((x, y); \mu_1(x, y)) \mid (x, y) \in X \times Y\},$$

$$R_2(y, z) = \{((y, z); \mu_2(y, z)) \mid (y, z) \in Y \times Z\}$$

yields a fuzzy set $R_{12}(x, z) = R_1(x, y) \circ_{MM} R_2(y, z)$ with

$$R_{12}(x, z) = \left\{ ((x, z);\ \max_{y} \min[\mu_1(x, y), \mu_2(y, z)]) \mid (x, y, z) \in X \times Y \times Z \right\}.$$

$R_1 \circ_{MM} R_2$ represents a fuzzy relation according to Definition 6.1 in the product space $X \times Z$, which means between values x and z.

If R_1 and R_2 can be represented as membership matrices, familiar methods for matrix multiplication can be used for max-min concatenation, where multiplication is replaced by minimum formation and summation by maximum formation.

Example 6.5 Let the two fuzzy relations $\mathbf{R_1}(x, y)$ and $\mathbf{R_2}(y, z)$ be given as subjective assessments comparing the crisp values of x, y, and z by the following membership matrices:

"x is greater than y" :	x \ y	1	2	5	10
	2	0.9	0.5	0.1	0
	5	1	0.9	0.5	0.1
	10	1	1	0.9	0.5

"y is approximately equal to z" :

y z	1	5	10	20
1	1	0.1	0	0
2	0.9	0.2	0	0
5	0.1	1	0.5	0.1
10	0	0.2	1	0.3

The numerical result $R(x, z) = R_1(x, y) \circ_{MM} R_2(y, z)$ of the max-min concatenation is calculated (according to Definition 6.4) as follows:

$\mu_{R_1}(y, z)$:

y z	1	5	10	20
1	1	0.1	0	0
2	0.9	0.2	0	0
5	0.1	1	0.5	0.1
10	0	0.2	1	0.3

x y	1	2	5	10
2	0.9	0.5	0.1	0
5	1	0.9	0.5	0.1
10	1	1	0.9	0.5

$\uparrow$

$\mu_{R_1}(x, y)$

x z	1	5	10	20
2	0.9	0.2	0.1	0.1
5	1	0.5	0.5	0.1
10	1	0.9	0.5	0.3

$\downarrow$

$\mu_R(x, z) = \mu_{R_1}(x, y) \circ_{MM} \mu_{R_2}(y, z)$

Similar to the case of concatenating sharp relations (according to Sect. 2.1), the result $R(x, z)$ can be interpreted as:

"x is greater than y and at the same time y is approximately equal to z."

The result is therefore a statement about the extent to which x is greater than z. If the physical units of x, y, and z are different, then this interpretation is to be understood as in Example 6.3.

In addition to max-min product, max-product ($\circ_{MP}$) is also used occasionally. When using the matrix multiplication schema , the formation of the inner product is retained, while the summation is replaced by the formation of the maximum. This calculates $\mathbf{R_1} \circ_{MP} \mathbf{R_2}$ as:

$$R_1 \circ_{MP} R_2 = \left\{ ((x, z); \max_y(\mu_{R_1}(x, y) \cdot \mu_{R_2}(y, z))) \,\middle|\, (x, y, z) \in X \times Y \times Z \right\}.$$

Another method is max-average concatenation ($\circ_{MA}$), in which the product formation is replaced by a mean value formation and the summation by the maximum formation. Then:

$$R_1 \circ_{\text{MA}} R_2 = \left\{ ((x, z); \ \frac{1}{2} \max_y (\mu_{R_1}(x, y) + \mu_{R_2}(y, z))) \mid (x, y, z) \in X \times Y \times Z \right\}.$$

If the numerical values for R_1 and R_2 are given (according to Example 6.5), the two membership matrices $\mu_{\text{MP}}(x, z)$ and $\mu_{\text{MA}}(x, z)$ are:

$\mu_{\text{MP}}(x, z)$: $\quad x \backslash z$	1	5	10	20
2	0.9	0.1	0.05	0.01
5	1	0.5	0.25	0.05
10	1	0.9	0.5	0.15

$\mu_{\text{MA}}(x, z)$: $\quad x \backslash z$	1	5	10	20
2	0.95	0.55	0.5	0.45
5	1	0.75	0.55	0.5
10	1	0.95	0.75	0.5

This example shows that the concatenation results can vary significantly depending on the concatenation method selected. The max-average method plays something of an outsider role, as the variability of the membership values is obviously much lower here than with the other two methods, making the set description less precise.

If the membership function $\mu(x, z)$ is interpreted as a possibility distribution and a worst-case scenario is to be considered, then the max-average method is preferable to the other two methods. As a rule, however, the max-min method is used, for example, for reasons of calculation speed.

Reference

Bothe, H.-H. (1995). *Bewertung mit unscharfen Mengen*. Technische Hochschule Ilmenau.

Chapter 7
Fuzzy Rule-Based Expert Systems

Abstract This chapter centers on fuzzy rule-based expert systems. These systems emulate human decision-making by combining expert knowledge with fuzzy logic to handle uncertainty and imprecise data. Such expert systems operate with degrees of truth, making them ideal for complex, real-world problems where boundaries are not clearly defined.

Rule-based expert systems model an expert who can handle the system to be operated rather than the system itself. According to Bothe (1995), this makes expert knowledge accessible to non-experts, who can then make decisions with the help of rule-based expert systems.

7.1 General Structure

In general, a number of requirements can be placed on rule-based expert systems:

- Communication with experts and users should take place in a *natural language* environment (standardized technical language).
- The rule-based expert system should make its decisions in a similar way to humans so that the decision-making process is easy to reconstruct and understand.
- Decision-making should also be guaranteed even if the input information is insufficient (i.e., the rule-based expert system's response to input conditions not stored in the knowledge base should be determined).
- There should be the possibility of expanding knowledge through learning.
- Subject-specific knowledge and problem-solving methods should be logically separated from each other.

A general block structure of a rule-based expert system is shown in Fig. 7.1. The actual rule-based expert system is marked with a dashed frame and "communicates" with both the expert and the user.

H.-H. Bothe, E. Portmann, *Computing with Words*, Fuzzy Management Methods,
https://doi.org/10.1007/978-3-032-24117-7_7

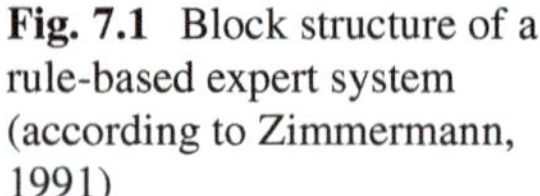

Fig. 7.1 Block structure of a rule-based expert system (according to Zimmermann, 1991)

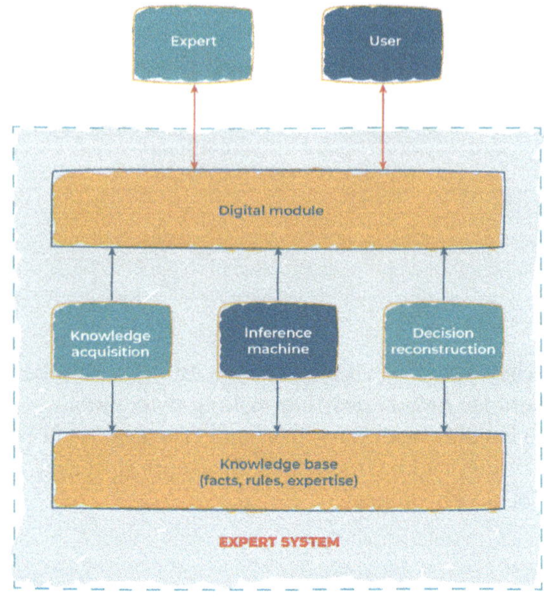

7.1.1 Dialogue Module

With the help of this module, the rule-based expert system can be prompted to "dialogue" by both the user who queries knowledge and the expert who generates knowledge.

7.1.2 Knowledge Base

The knowledge base forms the foundation of an rule-based expert system. It contains all of the available specialist knowledge. This is divided into declarative and procedural knowledge: Declarative knowledge describes the objects used by the expert (basic facts such as $\langle u_i = A_i \rangle$, concepts, meanings of terms, etc.) and their relationships to each other. It is stored in the database. Procedural knowledge describes the possibilities for drawing conclusions from the objects in the database using rules such as IF $u_i = A_i$ THEN $v_i = B_i$. These rules represent selected (and particularly important) precedents. They are stored in the rule base.

Example 7.1 In a medical rule-based expert system, various specialists can be regarded as experts in their specific field, whose expertise should be made available to the attending physician. The declarative knowledge consists, among other things, of the objects "possible diseases," "explanations for disease progression," and "disease symptoms, and symptoms when taking medication." Procedural knowledge consists of stored possible solutions and methods for drawing conclusions about the patient's causative diseases and their interaction based on recognized or suspected symptoms.

In rule-based expert systems, the rule base therefore consists of a series of "production rules" of the form

$$\text{IF } \langle \text{condition part} \rangle \text{ THEN } \langle \text{consequence part} \rangle.$$

We will see later how these production rules can be expressed with the help of fuzzy relations. Here are some basic characteristics:

– They are quasi-natural language.
– The knowledge base can be easily expanded by adding rules,
– The condition and consequence parts can be composed of several parts with the help of logical relationships.

To support experts and users, the dialogue module can include user-friendly rule editors and/or automatic input error checking.

There are also options for concatenated knowledge storage, which avoid the division into procedural and declarative knowledge.

7.1.3 Knowledge Acquisition

A separate module is usually integrated into the rule-based expert system for knowledge acquisition. It performs both guided expert questioning and, possibly, automatic learning strategies that build on the knowledge entered and generate new knowledge.

7.1.4 Inference Engine

Finally, the inference engine contains the methods for linking the individual rules of the knowledge base under given process conditions and drawing conclusions for specific target values.

Fuzzy rule-based expert systems are based on qualitative (fuzzy) expert statements and thus on (crisp) stored qualitative knowledge. In addition to the use of linguistic variables and operators, an essential concept is the method of "approximate reasoning." This will be presented in more detail in the following two chapters with the aid of fuzzy relations.

7.1.5 Decision Reconstruction

In order to make the decisions of the rule-based expert system verifiable for the expert and comprehensible for the user, a module for decision reconstruction can also be implemented.

7.2 Approximate Reasoning

A significant problem with rule-based systems—controllers or classifiers—is linking the current input values with the stored rules and using them to determine output values, to draw conclusions, or, in short, to "close." A solution often used for this task is the *modus ponens* of binary logic. Here, conclusions are drawn on the basis of an implication ("IF...THEN...rule" or "$\Rightarrow$") and a premise (value specification for the condition of the implication). The conclusions can be implemented quantitatively with the help of fuzzy relations. Modus ponens is a tautology (i.e., a logical relationship that is always true).

7.2.1 Modus Ponens

To illustrate the method, let us first assume two variables u and v with possible numerical values x_0 and y_0. The statements "$u = x_0$" and "$v = y_0$" (or "x_0" and "y_0" for short) are either true or false. A conclusion can only be drawn with certainty if the condition of implication is fulfilled by the premise. This results in the following schema :

$$
\begin{array}{ll}
\text{Implication:} & \text{IF } u = x_0 \text{ THEN } v = y_0, \\
\underline{\text{Premise:} \quad\quad u = x_0,} & \\
\text{Conclusion:} & v = y_0.
\end{array}
$$

or in shorthand notation:

$$[(x_0 \Rightarrow y_0) \wedge x_0] \rightarrow y_0. \tag{7.1}$$

The simple arrow ($\rightarrow$) is intended to indicate the process of inference but is otherwise treated as an implication. Modus ponens can be interpreted as follows: provided that the expert rule "$x_0 \Rightarrow y_0$" is true, the conclusion y_0 can be drawn when x_0 is given. However, the question arises as to what result v the rule should produce when given a general premise "$u = x_1$" with $x_1 \neq x_0$.

Furthermore, the truth value of the implication (i.e., the production rule) could also be questioned. In this case, (7.1) can be interpreted as follows: If the implication "$x_0 \Rightarrow y_0$" is true and the premise x_0 is also true, it can be concluded that y_0 is also true. Conversely, the truth value of the implication is only certainly false if $u = x_0$, but $v \neq y_0$; the other truth states are indeterminate. In order to resolve this "binary logical" conflict, the statement that is certainly false about the truth state is usually generalized according to an equivalence:

$$[x_0 \Rightarrow y_0] \Leftrightarrow [\neg(x_0 \wedge \neg y_0)] \Leftrightarrow [(\neg x_0 \vee y_0)]. \tag{7.2}$$

This results in the following truth table for the implication ($1 = $ true, $0 = $ false):

x_0	y_0	$x_0 \Rightarrow y_0$
0	0	1
0	1	1
1	0	0
1	1	1

The first two lines of the above table are incompatible with subjective logical perception, since a conclusion is drawn from a premise that is irrelevant to the implication. They are defined by postulating the validity of (7.2). A corresponding approach can only be applied to technical systems if extensive precautions are taken.

Comments

1. A truth table can be set up in a similar way for the conclusion (7.1).
2. In addition to the problem described, paradoxes also arise in Cartesian logic. For example, the truth value of the statement A "I always lie" cannot be clearly determined; for if A were true, "I" would also have to lie when formulating A, which would make A false, and vice versa.

7.2.2 Extended Modus Ponens

In contrast, approximate reasoning based on fuzzy rules and facts assumes an "extended modus ponens" based on gradual numerical truth values of the statements with possible intermediate values $v \in [0, 1]$ in the unit interval. To illustrate the method, two linguistic variables u and v with the linguistic values α and β and slightly modified values α' and β'' are assumed on the basic domains X and Y. The single and double quotation marks are intended to draw attention to the fact that α' and β'' result from α and β with different modification rules. These fuzzy values are mapped to the physical world with the help of fuzzy sets or numbers $\mathbf{A}(x)$,

$\mathbf{B}(y)$, $\mathbf{A}'(x)$, $\mathbf{B}''(y)$.[1] Given an implication and an associated premise, the following schema emerges:

$$\begin{array}{rl}
\text{Implication:} & \text{IF } u = \alpha \text{ THEN } v = \beta, \\
\hline
\text{Premise:} & u = \alpha', \\
\text{Conclusion:} & v = \beta''.
\end{array}$$

or in shorthand notation:

$$[(\alpha \Rightarrow \beta) \wedge \alpha'] \rightarrow \beta''. \tag{7.3}$$

The fuzziness of the inference β'' is due, on the one hand, to the fuzziness of α, β, and α' and, on the other hand, to the inference algorithm itself (i.e., to the conversion of the extended modus ponens into a mathematical formalism).

Example 7.2 A system is to be developed for automatically determining the taste of cherries from measured values, such as cherry color. According to expert opinion, *red cherries are sweet*:

$$\text{IF cherry color } u = \text{red THEN taste } v = \text{sweet.}$$

The terms *red* and *sweet* are mapped to a color and taste scale, respectively, using fuzzy sets $\mathbf{A}(x)$ and $\mathbf{B}(y)$. When a *very red* cherry is specified, the following basic situation arises:

$$\begin{array}{rl}
\text{Implication:} & \text{IF } u = \text{red THEN } v = \text{sweet,} \\
\hline
\text{Premise:} & u = \text{very red,} \\
\text{Conclusion:} & v = \text{more than sweet.}
\end{array}$$

The fuzziness of the conclusion "taste v = more than sweet" arises from the interaction of the specified fuzziness of the terms *red*, *sweet*, and *very red* with the fact that the condition "cherry color u = red" of the implication is only gradually fulfilled by the premise. The conclusion contains multiple fuzzinesses.

Without additional statements, it is easy to draw false conclusions in such a system (e.g., replace *sweet* with *tasty*; in this case, the cherry could already be too red to still taste good). The above example shows the need to create a rule base that describes as many basic states as possible, that is, a sufficient number of representative individual rules, in order to avoid false conclusions or instabilities in the process description. Creating such a rule base is one of the main problems when using fuzzy methods and cannot be generalized but must be carried out on

---

[1] The specification of arguments (x) or (y) for the fuzzy sets is intended to clarify the basic domain on which they are defined. With a sharp premise value, $u = x_0$, x_0 is interpreted as a singleton.

a problem-specific basis. As described, expert statements must be used for this purpose. Another possibility is for the developer to study the behavior of a system to be investigated by varying the parameters until he himself becomes an expert and is able to describe it in verbal statements.

7.3 Inference Using Fuzzy Relations

When a binary fuzzy relation R between two variables x, y and the variable value w (which is also fuzzy) is specified simultaneously, a relational equation is created that restricts the value range of the second variable value. The specified value can then be regarded as a crisp barrier or restriction for the second variable. As a consequence, a solution set is created whose characteristic function can itself be regarded as a restriction for its basic domain.

Example 7.3 Let us consider the binary crisp relation $R_>$: *greater than* between the variables x and y, given by $R_>(x, y) = \{(x, y) \in \mathbb{R}^2 \mid x > y\}$. When $x_0 = 18$ is specified, the range $B(y)$ of y is restricted to:

$$B(y) = \{y \in \mathbb{R} \mid y < 18\}.$$

Fuzzy relations establish fuzzy relationships between the variables involved. For example, the binary fuzzy relation $\mathbf{R}_\approx(x, y)$: *approximately equal* establishes a relationship between two linguistic values u and v regarding the extent to which the membership values of u at the possible locations x correspond to the membership values of v at the locations y. The specification of one linguistic variable value (e.g., of u) fuzzily restricts the value range of the other variable (e.g., of v). The fuzziness can arise both from the fuzziness of the relation and from the specification of a fuzzy value.

Definition 7.1 (Fuzzy Bound) Let $\mathbf{R}(x, y) \in \mathbf{P}(X \times Y)$ be a fuzzy relation between two variables u and v. Then, by assigning a crisp value from the basic domain X or a fuzzy set on X to u, a fuzzy bound $\mathbf{B}(y)$ is defined as an elastic boundary of the solution space of v. $\mathbf{B}(y)$ determines the degree of possibility that $y \in Y$ is a value of v if $x \in X$ is a value of u.

First, we will consider the special case where the uncertainty of the bound arises solely from the uncertainty of the relation when a crisp value is specified for u. Let $\mathbf{R} \in \mathbf{P}(X \times Y)$ be a fuzzy relation between two variables u and v. Then, assigning the value $u = x_0$ with $x_0 \in X$ creates a fuzzy barrier $\mathbf{B}(y) = \mathbf{R}(x_0, y)$ for the solution space of v. Conversely, this can also be interpreted as a fuzzy set $\mathbf{B}(y)$ being assigned to a crisp value x_0. With β as the linguistic equivalent of $\mathbf{B}(y)$, the fuzzy relation R can therefore be expressed at the point $u = x_0$ by a rule of the form:

$$\text{IF } u = x_0 \text{ THEN } v = \beta,$$

in which $\mathbf{R}$ does not explicitly appear.[2] However, this assignment is not unique, since identical fuzzy boundaries $\mathbf{B}(y)$ can arise when different values x_0 are specified. A fuzzy relation therefore clearly leads to an IF...THEN...rule, but not vice versa.

When $\mathbf{R}$ is explicitly specified, an input value $u = x_1 \neq x_0$ leads to the conclusion $v = \mathbf{B}''$.

Example 7.4 Given the numerical variables *input voltage u* and *output voltage v* of a system, the set $X = \{0, 1, 2, 3\}$ of possible input voltage values x, the set $Y = \{1, 2, 5, 10, 15\}$ of possible output voltage values y, and the fuzzy relation $\mathbf{R}_{\ll}(x, y)$: *significantly smaller than* with the membership function:

$\mu_{R_{\ll}}(x, y):$ $x\ y$	1	2	5	10	15
0	0.3	0.5	1	1	1
1	0.1	0.4	1	1	1
2	0	0.3	0.9	1	1
3	0	0.1	0.8	0.9	1

By specifying the crisp value $x_0 = 2$ for u, the fuzziness of $\mathbf{R}_{\ll}(x, y)$ restricts the value range of v in accordance with the row $[x = 2]$ of $\mathbf{R}_{\ll}(x, y)$. The fuzzy value $\mathbf{B}''(y)$ of v is thus (i.e., as a fuzzy bound for the value range of v):

$$\mathbf{B}''(y) = \{(y; \mu_{\beta''}(y)) \mid (1; 0),\ (2; 0.3),\ (5; 0.9),\ (10; 1),\ (15; 1)\}.$$

The conclusion $v = \mathbf{B}''(y)$ represents the fuzzy output value of a fuzzy system whose transfer behavior is characterized by the fuzzy relation $\mathbf{R}_{\ll}(x, y)$ for the crisp input value x_0.

Remarks

1. In Example 7.4, a situation similar to the rule evaluation in Example 1.7 arises, but the fuzzy output variable is now calculated from the crisp input variable using a fuzzy relation instead of an IF...THEN...rule. This raises the question of whether and how these rules can be transformed into fuzzy relations in order to mathematically justify the heuristic approach in Example 1.7.
2. If conclusions $\mathbf{B}_{x_0}(y)$ are known for all $x_0 \in X$, a corresponding fuzzy relation $\mathbf{R}(x, y)$ on $X \times Y$ can be constructed exactly.
3. The membership function $\mu_B(y)$ of each fuzzy set $\mathbf{B}_{(y)} \in \mathbf{P}(Y)$ can always be regarded—similar to a filter function—as a fuzzy bound for the value range of a variable v whose variability domain matches the entire basic domain Y.

---

[2] In applications such as fuzzy controllers, IF...THEN...production rules are specified, and the task is to find the corresponding fuzzy relations $\mathbf{R}$.

A generalized inference arises when the input value for u is also fuzzy. Let $\mathbf{R} \in \mathbf{P}(X \times Y)$ be a given fuzzy relation between u and v, and let $\mathbf{A}' \in \mathbf{P}(X)$ be the fuzzy value of u. The resulting inference $v = \mathbf{B}''(y)$ is then determined in a doubly fuzzy manner by $\mathbf{A}'(x)$ and $\mathbf{R}(x, y)$.

Zadeh (1973) points out the possibility of interpreting each fuzzy set $\mathbf{A}'(x)$ as a single-valued fuzzy relation on X and proposes the max-min concatenation $\mathbf{A}'(x) \circ_{\mathrm{MM}} \mathbf{R}(x, y)$ according to Definition 6.4 for calculating the resulting double fuzzy bound $\mathbf{B}''(y)$. The membership function $\mu_{B''}(y)$ of $\mathbf{B}''(y)$ is then given by:

$$\mu_{B''}(y) = \max_{x} \min[\mu_{A'}(x), \mu_R(x, y)]. \tag{7.4}$$

Example 7.5 Let the input voltage be u; the output voltage be v; the basic domain be $X = \{1, 2, 3, 4\}$ for the possible input and output voltage values $x, y \in X$; and the transfer behavior of a system in the form of a fuzzy relation $\mathbf{R}$:   "u is approximately equal to v" be given as a matrix as follows:

$\mu_R(x, y) : x \backslash y$	1	2	3	4
1	1	0.4	0	0
2	0.4	1	0.4	0
3	0	0.4	1	0.4
4	0	0	0.4	1

Given a fuzzy input voltage $u = $ small, expressed by the fuzzy set

$$\mathbf{A}' = \{(1; 1),\ (2; 0.6),\ (3; 0.2),\ (4; 0)\},$$

the calculation of the membership values of $\mathbf{B}''(y)$ can be represented in matrix notation.

$x \backslash y$	1	2	3
1	1	0.4	0
2	0.4	1	0.4
3	0	0.4	1
4	0	0	0.4

$\mathbf{R} \rightarrow$

x	1	0.6	0.2	0

$\uparrow$

$\mathbf{A}'$

y	1	0.6	0.4	0.2

$\downarrow$

$\mathbf{B}'' = \mathbf{A} \circ_{\mathrm{MM}} \mathbf{R}$

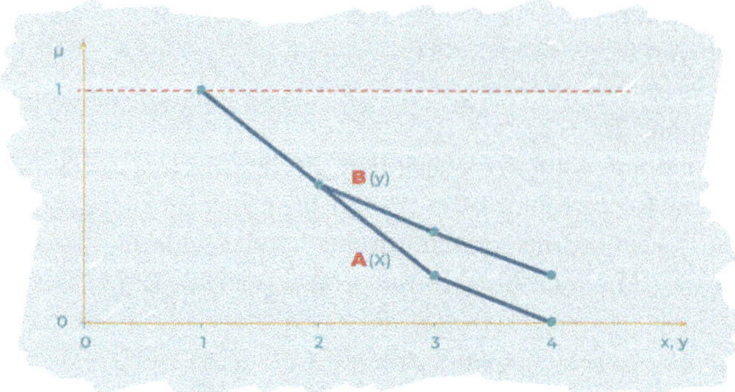

Fig. 7.2 Membership functions of input set $\mathbf{A}'$ and output set $\mathbf{B}''$

The fuzzy barrier $\mathbf{B}''$ for the possible values y of the output voltage v is given by (Fig. 7.2)

$$\mathbf{B}'' = \{(1;\, 1),\ (2;\, 0.6),\ (3;\, 0.4),\ (4;\, 0.2)\}.$$

This result can be interpreted as follows:

$$\mathbf{R}(x, y) : \quad u \text{ and } v \text{ are approximately the same magnitude,}$$

$$\mathbf{A}'(x) : \quad u \text{ is small,}$$

$$\overline{\mathbf{B}''(y) : \quad v \text{ is fairly small.}}$$

In addition, the other operations described in Sect. 6.2 can also be used to concatenate fuzzy relations. A generalized approach to reasoning with fuzzy statements can be proposed as follows, based on Delgado (1990). To calculate $\mathbf{B}''(y)$ given $\mathbf{A}'(x)$, the following steps must be performed according to this calculation rule:

1. The linguistic terms α and α' are expressed as fuzzy sets $\mathbf{A}$ and $\mathbf{A}'$ on the basic domain X, β as a fuzzy set $\mathbf{B}$ on the base domain Y ($\mathbf{B}''$ is initially unknown).
2. An implication link $\mathbf{I}(\mathbf{A}, \mathbf{B})$ is selected which, for all $(x, y) \in X \times Y$, converts the rule $\alpha \rightarrow \beta$ into a fuzzy relation $\mathbf{R}_I(x, y)$:

$$\mathbf{R}_I(x, y) = \mathbf{I}[\mathbf{A}(x), \mathbf{B}(y)]. \tag{7.5}$$

3. $\mathbf{R}(x, y)$ is converted into a relation $\mathbf{R}_T(x, y)$, taking into account $\mathbf{A}'(x)$ and under the condition "for all $x \in X$ and $y \in Y$, $\mathbf{A}'(x)$ and $\mathbf{R}(x, y)$ shall apply":

$$\mathbf{R}_T(x, y) = \mathbf{A}'(x) \,\square\, \mathbf{R}_{I(x,y)}, \tag{7.6}$$

where $\square$ on $Y \in Y$ is a representation of *and* with respect to the x values.

4. $\mathbf{B}''(y)$ is determined for all $y \in Y$ according to the following formula:

$$\mathbf{B}''(y) = \mathrm{hgt}_{x \in X}[\mathbf{R}_T(x, y)]. \tag{7.7}$$

With (7.5) to (7.7), the rule $\alpha \to \beta$ generates the fuzzy boundary $\mathbf{B}''(y)$ for the possible values of v when α' is specified, with:

$$\mathbf{B}''(y) = \mathrm{hgt}_{x \in X}\big[\mathbf{A}'(x) \,\square\, \mathbf{I}[\mathbf{A}(x), \mathbf{B}(y)]\big].$$

The requirement $\mathbf{B}''(y) \subseteq \mathbf{B}(y)$ is now imposed on $\mathbf{B}''(y)$, which means that no conclusion can be drawn for those values of the basic domain that are not covered by the rule; this means $\mu_{B''}(y) \leq \mu_B(y)$.

To implement generalized modus ponens, two decisions must be made: first, which implication relation to use and, second, which operator ($\square$) to use to represent the *and*; the choice of ($\square$) should be made depending on the implication relation chosen. Numerous possible combinations are discussed (e.g., in Tilli, 1991). Some interesting variants are presented below.

7.3.1   Case According to Mamdani (1975)

A very simple implication relation R_M uses the Cartesian product $\otimes$ and was proposed by Mamdani; it is calculated as:

$$\mathbf{R}_M = \mathbf{A} \otimes \mathbf{B}. \tag{7.8}$$

This results in the membership function $\mu_M(x, y)$ for:

$$\mu_M(x, y) = \min[\mu_A(x), \mu_B(y)]. \tag{7.9}$$

In this case, the intersection is used as the representation of *and* in (7.6); the conclusion $\mathbf{B}''(y)$ is then calculated according to (7.6) by max-min concatenation of $\mathbf{A}'$ and $\mathbf{R}_M$. With (7.5), (7.6), and (7.7), the membership function $\mu_{MB''}(y)$ is created with:

$$
\begin{aligned}
\mu_{MB''}(y) &= \sup_{x \in X} \min[\mu_{A'}(x), \mu_M(x, y)] \\
&= \sup_{x \in X} \min[\mu_{A'}(x), \min(\mu_A(x), \mu_B(y))] \\
&= \sup_{x \in X} \min[\min(\mu_{A'}(x), \mu_A(x)), \mu_B(y)] \quad (7.10) \\
&= \sup_{x \in X} \min[\mu_{A' \cap A}(x), \mu_B(y)] \\
&= \min\big[\sup_{x \in X} \mu_{A' \cap A}(x), \mu_B(y)\big] \leq \mu_B(y).
\end{aligned}
$$

For $\beta_M = \sup_{x \in X} \mu_{A' \cap A}(x) = \mathrm{hgt}(\mathbf{A}' \cap \mathbf{A}) = 1$, if the condition part in (7.5) is completely fulfilled, the conclusion $\mathbf{B}'' = \mathbf{B}$; otherwise, with $\beta_M \in [0, 1]$:

$$\mu_{MB''}(y) = \min[\beta_M, \mu_B(y)]. \tag{7.11}$$

The factor β_M is also called the *degree of activation* of the rule. Formula (7.11) thus describes the max-min inference method, as already presented in Example 1.7 using a simple controller behavior. In that example, however, there were two rules whose fuzzy subconclusions were then linked to the final conclusion using the union operator. If the min operator in (7.10) and (7.11) is replaced by the algebraic product, the max-product (or max-dot) inference method is created (see Sect. 8.2).

Example 7.6 An expert gives a production rule IF $u = \alpha$ THEN $v = \beta$, which represents a consequence $v = \mathbf{B}$ when the value $u = \mathbf{A}$ is specified. Let u and v be the two linguistic variables and

$$\mathbf{A} = \{(1; 0.3), (2; 0.6), (3; 1), (4; 0)\}, \quad \mathbf{B} = \{(1; 0.2), (2; 1), (3; 0.6), (4; 0.3)\}.$$

We are looking for the fuzzy value $\mathbf{B}''$ of v for the value specification of u with:

$$\mathbf{A}' = \{(1; 1), (2; 0.5), (3; 0.4), (4; 0.1)\}.$$

With the implication representation $\mathbf{I}(\mathbf{A}, \mathbf{B}) = \mathbf{A} \otimes \mathbf{B}$, we get $\mu_{R_I}(x, y) = \min[\mu_A(x), \mu_B(y)]$:

$\mu_{R_I}(x, y) : x \backslash y$	1	2	3	4
1	0.2	0.3	0.3	0.3
2	0.2	0.6	0.6	0.3
3	0.2	1	0.6	0.3
4	0	0	0	0

The membership function of $\mathbf{R}_T(x, y)$ is calculated using the min operator with $\mu_{RT}(x, y) = \min_{x \in X}[\mu_{A'}(x), \mu_{RI}(x, y)]$, together with:

$\mu_{R_T}(x, y) : x \backslash y$	1	2	3	4
1	0.2	0.3	0.3	0.3
2	0.2	0.5	0.5	0.3
3	0.2	0.4	0.4	0.3
4	0	0	0	0

This results in $\mathbf{B}''(y) = \sup_{x \in X} \mathbf{R}_T(x, y)$ as:

$$\mathbf{B}''(y) = \{(1; 0.2), (2; 0.5), (3; 0.5), (4; 0.3)\}.$$

7.3.2 *Case According to Kleene-Dienes (see Trillas, 1992)*

From the equivalence $[A \Rightarrow B] \Leftrightarrow [A^c \vee B]$ of binary logic, an alternative implication relation $\mathbf{R}_{KD}$ arises by analogy. Since the simple application of union requires an identical basic domain, the formation of a Cartesian product is also necessary here. Then, as an extension of (7.8), we obtain:

$$\mathbf{R}_{KD} = (\mathbf{A}^c \otimes Y) \cup (X \otimes \mathbf{B}). \tag{7.12}$$

This calculates the membership function $\mu_{KD}(x, y)$ to:

$$\begin{aligned}
\mu_{KD}(x, y) &= \max[\min(1 - \mu_A(x), 1),\ \min(1, \mu_B(y))] \\
&= \max[1 - \mu_A(x),\ \mu_B(y)].
\end{aligned} \tag{7.13}$$

If, as in (7.9), the restricted product is used here to represent *and* in (7.6), the result is, with $\sup_{x \in X} \mu_{A'}(x) = 1$ (i.e., normalized $\mathbf{A}'$), for the conclusion $\mu_{KD\beta''}(y)$:

$$\begin{aligned}
\mu_{KD\beta''}(y) &= \sup_{x \in X} \max[0,\ \mu_{A'}(x) + \max(1 - \mu_A(x), \mu_B(y)) - 1] \\
&= \sup_{x \in X} \max\big[0,\ \max(\mu_{A'}(x) - \mu_A(x),\ \mu_{A'}(x) + \mu_B(y) - 1)\big] \\
&= \max\left[\sup_{x \in X}(\mu_{A'}(x) - \mu_A(x)),\ \sup_{x \in X}(\mu_{A'}(x) + \mu_B(y) - 1)\right] \\
&= \max\left[\sup_{x \in X}(\mu_{A'}(x) - \mu_A(x)),\ \mu_B(y)\right].
\end{aligned}$$

With $\beta_{KD} = \sup_{x \in X}[\mu_{A'}(x) - \mu_A(x)]$, it follows that:

$$\mu_{KDB''}(y) = \max[\beta_{KD}, \mu_B(y)]. \tag{7.14}$$

In the case of $\sup_{x \in X}[\mu_{A'}(x) - \mu_A(x)] \leq \mu_B(y)$ (i.e., for sufficient agreement between premises $\mathbf{A}$ and $\mathbf{A}'$), the following applies to the conclusion $\mu_{KDB''}(y) = \mu_B(y)$ for all $y \in Y$, i.e., $\mathbf{B}'' = \mathbf{B}$. This can be clearly interpreted as meaning that, in extension of (7.11), a certain range of validity for the rule is created.

7.3.3 *Case According to Reichenbach*

If the union in (7.12) is replaced by the arithmetic sum, the following implication relation $\mathbf{R}_R$ can also be derived by analogy:

$$\mathbf{R}_R = (\mathbf{A}^c \otimes Y) + (X \otimes \mathbf{B}). \tag{7.15}$$

This is known as the *Reichenbach implication* and has the membership function:

$$
\begin{aligned}
\mu_R(x, y) &= \min[1 - \mu_A(x), 1] + \min[1, \mu_B(y)] \\
&\quad - \min[1 - \mu_A(x), 1] \cdot \min[1, \mu_B(y)] \\
&= 1 - \mu_A(x) + \mu_B(y) - (1 - \mu_A(x))\mu_B(y) \\
&= 1 - \mu_A(x) + \mu_A(x)\mu_B(x).
\end{aligned}
\tag{7.16}
$$

In this case, the restricted product is used as a representation of *and*; the conclusion $\mathbf{B}''(y)$ is then calculated using (7.5), (7.6), and (7.7) as follows:

$$
\begin{aligned}
\mu_{RB''}(y) &= \sup_{x \in X} \max[0, \ \mu_{A'}(x) - \mu_A(x) + \mu_A(x).\mu_B(y)] \\
&= \max\left[0, \ \sup_{x \in X}\left(\mu_{A'}(x) - \mu_A(x) + \mu_A(x).\mu_B(y)\right)\right].
\end{aligned}
$$

If $\sup_{x \in X}[\mu_{A'}(x)] = 1$, $\mathbf{A}$ is normalized, the following applies:

$$\mu_{RB''}(y) \le \max\left[0, \ \sup_{x \in X}\left[\mu_{A'}(x) - \mu_A(x)\right] + \mu_B(y)\right].$$

With $\beta_R = \sup_{x \in X}\left[\mu_{A'}(x) - \mu_A(x)\right]$, this then results in:

$$\mu_{RB''}(y) \le \max[0, \ \beta_R + \mu_B(y)]. \tag{7.17}$$

The requirement $\forall y \in Y : \mu_{RB''}(y) \le \mu_B(y)$ is only reliably fulfilled if $\forall x \in X : \beta_R \le 0$. In practice, this cannot be maintained for arbitrary $\mathbf{A}'(x)$.

7.3.4   Case According to Łukasiewicz

If the union in (7.12) is replaced by the restricted sum, the following implication relation $\mathbf{R}_L$ can be derived by analogy:

$$\mathbf{R}_L = (\mathbf{A}^c \otimes Y) \cup (X \otimes \mathbf{B}). \tag{7.18}$$

The membership function is then calculated as:

$$\mu_L(x, y) = \min[1, \ 1 - \mu_A(x) + \mu_B(y)]. \tag{7.19}$$

In this case, the restricted product is used as a representation of *and* in (7.6); the conclusion $\mathbf{B}''(y)$ is then calculated using (7.5), (7.6), and (7.7) as well as $\sup_{x \in X} \mu_{A'}(x) = 1$ as follows:

$$\mu_{LB''}(y) = \sup_{x \in X} \max[0, \ \mu_{A'}(x) + \min(1, \ 1 - \mu_A(x) + \mu_B(y)) - 1]$$

$$= \sup_{x \in X} \max[0, \ \min[\mu_{A'}(x), \ \mu_{A'}(x) - \mu_A(x) + \mu_B(y)]]$$

$$= \max\left[0, \ \min\left(\sup_{x \in X} \mu_{A'}(x), \ \sup_{x \in X}\left(\mu_{A'}(x) - \mu_A(x)\right) + \mu_B(y)\right)\right]$$

$$= \max\left[0, \ \min\left(1, \ \sup_{x \in X}\left(\mu_{A'}(x) - \mu_A(x)\right) + \mu_B(y)\right)\right].$$

With $\beta_L = \sup_{x \in X}\left[\mu_{A'}(x) - \mu_A(x)\right]$, the fuzzy conclusion is finally calculated as:

$$\mu_{LB''}(y) = \max[0, \ \min(1, \ \beta_L + \mu_B(y))]. \tag{7.20}$$

The requirement $\forall y \in Y : \mu_{LB''}(y) \leq \mu_B(y)$ is only definitively fulfilled for (7.20) if $\forall x \in X : \beta_L \leq 0$. As in (7.17), the corresponding requirement $\forall x \in X : \mu_{A'}(x) \leq \mu_A(x)$ is difficult to meet in practice.

7.3.5   Case According to Zadeh

The Mamdani implication relation is a simplification of the Zadeh implication relation $\mathbf{R}_Z$, which is defined as follows:

$$\mathbf{R}_Z = (\mathbf{A} \otimes \mathbf{B}) \cup (\mathbf{A}^c \otimes Y). \tag{7.21}$$

The interpretation of (7.21) is:

$$\text{IF } u = \mathbf{A} \text{ THEN } v = \mathbf{B} \text{ ELSE } v = Y.$$

This allows us to calculate the membership function $\mu_Z(x, y)$ as:

$$\mu_Z(x, y) = \max[\min[\mu_A(x), \mu_B(y)], \ \min[1 - \mu_A(x), 1]]$$
$$= \max[1 - \mu_A(x), \ \min[\mu_A(x), \mu_B(y)]]. \tag{7.22}$$

The Zadeh implication is used together with the restricted product. The formula for determining the conclusion can be calculated analogously to the above cases. According to Tilli (1991), in certain situations, conclusions arise that do not correlate with logical intuition. For this reason, the Zadeh implication will not be pursued further here.

7.3.6 *Case Gödel Implication*

In addition to the empirically based implication formulas, the Gödel implication relation $\mathbf{R}_G$ is also specified here, the properties of which are described in more detail in Sect. 7.4. It is used in connection with the intersection. The membership function $\mu_G(x, y)$ used is:

$$\mu_G(x, y) = \begin{cases} 1 & \text{for} \quad \mu_A(x) \leq \mu_B(y), \\ \mu_B(y), & \text{otherwise.} \end{cases} \tag{7.23}$$

The membership function of the conclusion $\mathbf{B}''$ is thus:

$$\mu_{GB''}(y) = \sup_{x \in X} \min \left[\mu_{A'}(x), \begin{cases} 1 & \text{for} \quad \mu_A(x) \leq \mu_B(y), \\ \mu_B(y), & \text{otherwise} \end{cases} \right] \tag{7.24}$$

Approximate reasoning using Gödel's implication is described in more detail for several given rules in Sect. 7.4. Other implication relations can be found, for example, in Mizumoto and Zimmermann (1982) and Tilli (1991). Section 8.2 also lists two inference methods that circumvent the determination of fuzzy relations and are used in controller design .

Another method of inference consists of using the extension principle to transfer the modus ponens of a multivalued Łukasiewicz logic to terms of the linguistic variable "truth." This approach is useful if the numerical truth value v of the implication is gradually given with $v \in [0, 1]$. Methods of inference can also be applied that use modus tollens, syllogism, or the contraposition of binary logic instead of modus ponens:

$$\text{Modus tollens:} \quad [(A \Rightarrow B) \wedge \neg B] \to \neg A, \tag{7.25}$$

$$\text{Syllogism:} \quad [(A \Rightarrow B) \wedge (B \Rightarrow C)] \to (A \Rightarrow C), \tag{7.26}$$

$$\text{Contraposition:} \quad (A \Rightarrow B) \to (\neg B \Rightarrow \neg A). \tag{7.27}$$

These methods of approximate reasoning lead to different results in each case; the specific choice is based more on heuristics, and there is no uniform theoretical basis. For a more detailed discussion, see Baldwin (1979), Mizumoto and Zimmermann (1982), and Zimmermann (1991).

If several rules for describing the system behavior are specified in a fuzzy rule-based expert system controller or classifier, a system of relational equations arises that must be solved. Two fundamentally different methods are used for this purpose: either a uniform transfer relation $\mathbf{R}(x, y)$ is sought that determines the entire system, or, given an input value $u = \mathbf{A}'(x)$, partial conclusions $v = \mathbf{B}_{i''}(y)$ are first calculated for the individual rules (i), which are then linked to form the conclusion $\mathbf{B}(y)$.

The facts of the conclusion based on individual inferences can be illustrated with the aid of a schema that represents an extension of the illustration in Example 1.7 (cooling valve): the specification of an originally crisp input value is to be replaced by that of a fuzzy value. This changes the "activation degree" β_i of the current rule, which means a corresponding change in the output quantity of this rule. When several rules are taken into account, the crisp output value also changes. Such an example is presented in Sect. 7.4.

7.3.7 Trustworthy

In practical cases, the rule base consists of 5–100 rules, which may well originate from different experts. This immediately implies that the individual rules may have different "confidence values." Even if only a single expert is consulted, the rule evaluation should take into account, for example, the influence of varying pronounced perceptions of visual and acoustic stimuli on the rule formulation. For this reason, a confidence value q_i should be introduced for each rule for "inverse weighting"; the easiest way to do this is to modify the current activation degrees β_i according to which the new activation degrees β_i^* are calculated as follows:

$$\beta_i^* = q_i \beta_i. \tag{7.28}$$

The introduction of confidence values immediately raises the new task of quantifying them. In practical terms, this leads to the question of how the contributions of the individual rules to the overall result should be weighted. This assessment is obviously very problem-specific and should be carried out very carefully by the developer.

In simple rule-based expert systems such as controllers or classifiers, their signal transmission behavior can also be optimized by varying the confidence values. It is also possible to set the confidence values in an automatic learning process. The use of neural networks for this purpose is described by Wakami (1992). In the following chapters, however, despite its importance, this modification of the rule activation degrees will be omitted in favor of a clear description of the basic processes.

The developer must check whether the methods used lead to reasonable conclusions, which means that it must be determined whether the system can adequately replace human experts in their conclusions or decision-making. In complex systems, this cannot always be fully verified. Therefore, machine-based expert systems should only be used in cases where they are embedded in redundant decision-making processes and the final decisions are made by human experts (on this topic; cf. Puppe, 1988).

7.4 Fuzzy Relational Equation Systems

In most cases, the fuzzy rule-based expert system will have several input variables $u_1, \ldots, u_m$ and output variables $v_1, \ldots, v_k$. These can be summarized as the input vectors $\mathbf{u} = (u_1, \ldots, u_m)$ and $\mathbf{v} = (v_1, \ldots, v_k)$.[3] In order to clarify the fundamental relationships, we will assume scalar variables u and v. Let us consider a fuzzy rule with an initially unknown fuzzy relation $\mathbf{R}$ according to (7.5) and an operation $\circ_T$ according to (7.6):

$$\text{IF } u = \mathbf{A} \text{ THEN } v = \mathbf{B} \quad \Rightarrow \quad \mathbf{B} = \mathbf{A} \circ_T \mathbf{R}. \tag{7.29}$$

The expert knowledge will now be available in the form of a rule base containing a large number of individual rules. Let us assume a rule base:

$$\text{IF } u = \mathbf{A}_i \text{ THEN } v = \mathbf{B}_i, \tag{7.30}$$

with n individual production rules. It too should respond to any input values with a specific output value. It therefore makes sense to describe the transfer behavior of the rule-based expert system using a single, rule base-specific fuzzy relation R. This means that the following system of relational equations must be solved:

$$\mathbf{B}_i = A_i \circ_T \mathbf{R}, \quad i = 1, \ldots, n. \tag{7.31}$$

The premises of the rules may contain only intersection operators; union operators are eliminated by creating additional rules. The operators presented in Sect. 6.2 for concatenating fuzzy relations are conceivable as linking operators ($\circ_T$).

Each production rule according to (7.31) is understood as a value assignment $u = \mathbf{A}_i$, which is caused by a value assignment $v = \mathbf{B}_i$ and through which the fuzzy relation R (constituting the fuzzy rule-based expert system) is effected. $\mathbf{R}$ thus describes how the fuzzy rule-based expert system should respond to the input values defined in the rule base and represents a kind of fuzzy transfer function that is specified by the expert as a rule base according to (7.30). The implication operators presented in Sect. 7.3 are used to calculate $\mathbf{R}$. It is conceivable to either postulate a common relation $\mathbf{R}$ at the outset or to first establish a separate relation $\mathbf{R}_i$ for each individual rule and then combine these to form a rule base-specific relation $\mathbf{R}$. The exact procedure is explained in Sect. 8.2.

In addition, the rule-based expert system should of course also respond to input values $\mathbf{A}'$ that are not explicitly specified. This is then set according to (7.4) as:

$$v = \mathbf{A}' \circ_T \mathbf{R}. \tag{7.32}$$

---

[3] Only intersection operators may be present in the premises; union operators are to be eliminated by creating additional rules.

The relational equation system (7.32) cannot always be solved in the sense that a common fuzzy relation $\mathbf{R}$ exists for the operation $\circ_T$. Such a case can occur, for example, when several production rules contradict each other. The aim must then be to minimize the errors that arise in particularly important statements or states. For this purpose, the rules can, for example, be organized into groups, and separate fuzzy relations can be calculated for these groups. This results in a system of several inference engines working in parallel.

Some of the fuzzy output values can, under certain circumstances, be combined to form new, common fuzzy values and fed into the inference engine as additional input variables. This process corresponds to an extension of switching networks to switching circuits in digital technology, as described (e.g., described by Giloi & Liebig, 1980). It can serve as the basis for the construction of memory-based state controllers.

First, let us assume a fuzzy relation $\mathbf{R}$ according to (7.5) and an operation $\circ_T$ according to (7.6):

$$\text{IF } u = \mathbf{A}_i \text{ THEN } v = \mathbf{B}_i \;\Rightarrow\; \mathbf{B}_i = \mathbf{A}_i \circ_T \mathbf{R}. \tag{7.33}$$

To solve a single relational equation, one of the implication relations described in Sect. 7.3 can be used. Since (7.33) is a relational equation system, there are also several possible fuzzy relations $\mathbf{R}'$, $\mathbf{R}''$, ... instead of a unique solution. According to Bandemer (1990), it can be shown for $(\circ_T) = (\circ_{\text{MM}})$ that the solution set $\mathcal{R} = \{\mathbf{R} \in \mathbf{P}(X \times Y) \mid \mathbf{B} = \mathbf{A} \circ_{\text{MM}} \mathbf{R}\}$ is an *upper semilattice* with respect to the inclusion of fuzzy sets (i.e., with two solutions $\mathbf{R}'$, $\mathbf{R}''$, the union $\mathbf{R}' \cup \mathbf{R}''$ is also a solution of (7.33)):

$$\mathbf{R}', \mathbf{R}'' \in \mathcal{R} \Rightarrow \mathbf{R}' \cup \mathbf{R}'' \notin \mathcal{R}. \tag{7.34}$$

$\mathbf{R}' \cup \mathbf{R}''$ is the smallest solution of (7.33) with respect to inclusion that contains $\mathbf{R}'$ and $\mathbf{R}''$. Therefore, the solution set $\mathcal{R}$ can contain at most one maximum element with respect to $\subseteq$—the largest solution of (7.33). The calculation of an optimal fuzzy implication relation for solving the fuzzy relational equation system is reduced to this largest solution.

Simplifying approaches that use, for example, the Cartesian product and are employed in particular in the field of controller design are explained in more detail in Sect. 8.2. For the optimal solution of (7.33), a new set-algebraic operation is to be defined based on (7.23).

Definition 7.2 (Gödel Operation $\star$) Let $\mathbf{A} \in \mathbf{P}(X)$ and $\mathbf{B} \in \mathbf{P}(Y)$. Then the operation $\mathbf{A} \star \mathbf{B} \in \mathbf{P}(X \times Y)$ is defined by:

$$\mu_{A \star B}(x, y) = \begin{cases} 1 & \mu_A(x) \leq \mu_B(y), \\ & \text{for} \\ \mu_B(y), & \mu_A(x) > \mu_B(y). \end{cases}$$

Theorem 7.1 *If $\mathbf{R} = \mathbf{A} \star \mathbf{B}$ is a solution of $\mathbf{B} = \mathbf{A} \circ_{\mathrm{MM}} \mathbf{R}$, then it is also the largest solution with respect to inclusion. Otherwise, $\mathbf{B} = \mathbf{A} \circ_{\mathrm{MM}} \mathbf{R}$ is unsolvable (Gottwald, 1984).*

Example 7.7 Let $\mathbf{A} = \{(1; 0.9), (2; 1), (3; 0.7)\}$, $\quad \mathbf{B} = \{(1; 1), (2; 0.4), (3; 0.8), (4; 0.7)\}$. With regard to inclusion, the largest solution $R_{\max}$ for $B = \mathbf{A} \circ_{\mathrm{MM}} \mathbf{R}$ is given by Theorem 7.1 as $R_{\max} = \mathbf{A} \star \mathbf{B}$ with:

$\mu_{R_{\max}}(x, y) : x \;\; y$	1	2	3	4
1	1	0.4	0.8	0.7
2	1	0.4	0.8	0.7
3	1	0.4	1	1

Another solution is given, for example, by $\mathbf{R}^*$ with:

$\mu_{R^*}(x, y) : x \;\; y$	1	2	3	4
1	0	0.4	0.8	0
2	1	0	0	0
3	0	0	0	0.7

A sample yields $\mathbf{B} = \mathbf{A} \circ_{\mathrm{MM}} \mathbf{R}_{\max} = \mathbf{A} \circ_{\mathrm{MM}} \mathbf{R}^*$ in both cases.

The expert knowledge will now be available in the form of a rule base containing a large number of individual rules. The relational equation system cannot be solved in all cases in the sense that a common fuzzy relation $\mathbf{R}$ exists for the operation $\circ_T$. Such a case can occur, for example, when several production rules contradict each other. The aim must then be to minimize the errors that arise for particularly important statements or states. In the case $(\circ_T) = (\circ_{\mathrm{MM}})$, an optimal solution with regard to inclusion can be specified as described in Theorem 7.2.

Theorem 7.2 *A system of relational equations $\mathbf{B}_i = A_i \circ_{\mathrm{MM}} \mathbf{R}$ can be solved according to $\mathbf{R}$ if and only if*

$$\mathbf{C} = \bigcap_{i=1\ldots n} (\mathbf{A}_i \star \mathbf{B}_i)$$

is a solution. Then $\mathbf{C}$ is simultaneously the largest solution with respect to inclusion. Then $\mathbf{C}$ is simultaneously the largest solution with respect to inclusion for $\mathbf{R}$. If no solution exists, $\mathbf{C}$ is a particularly favorable approximate solution (Bandemer, 1990).

Example 7.8 A fuzzy controller is to regulate the *dye addition* v to a liquid depending on its *flow velocity* u. The following rule base is given:

1. IF $u = $ fast THEN $v = $ large amount.
2. IF $u = $ slow THEN $v = $ small amount,

where $U = \{1, 2, 3\}$ and $V = \{10, 20, 30\}$ are the basic domains for the linguistic variables u and v, $x \in U$ and $y \in V$ the elements. The terms are mapped to U and V with:

$$fast : A_1 = \{(1; 0), (2; 0.5), (3; 1)\},$$

$$slow : A_2 = \{(1; 1), (2; 0.4), (3; 0)\},$$

$$large\ amount : B_1 = \{(10; 0), (20; 0.6), (30; 1)\},$$

$$small\ amount : B_2 = \{(10; 1), (20; 0.3), (30; 0)\}.$$

How large is the fuzzy output value $\mathbf{B}''$ of the controller when a fuzzy input value $\mathbf{A}' = \{(1; 0.1), (2; 1), (3; 0)\}$ is present? To answer this question, the rule base is represented by a fuzzy relational equation system $\mathbf{B}_i = A_i \circ \mathbf{R}$. Applying the Gödel operation results in:

$$\mathbf{B}_1 = \mathbf{A}_1 \circ_{MM} \mathbf{R},$$

$$\mathbf{B}_2 = \mathbf{A}_2 \circ_{MM} \mathbf{R}.$$

To solve for $\mathbf{R}$, the fuzzy relations to be linked, $\mathbf{C}_1 = A_1 \star B_1$, and $\mathbf{C}_2 = A_2 \star B_2$, are first calculated:

$\mu_{C1}(x, y):$ xy	10	20	30
1	1	1	1
2	0	1	1
3	0	0.6	1

$\mu_{C1}(x, y):$ xy	10	20	30
1	1	0.3	0
2	1	0.3	0
3	1	1	1

The largest solution $\mathbf{R}$ with respect to inclusion is then calculated according to

$$\mathbf{R} = C_1 \cap C_2,$$

with the following membership matrix $\mu_R(x, y)$:

$\mu_R(x, y):$ xy	10	20	30
1	1	0.3	0
2	0	0.3	0
3	0	0.6	1

The fuzzy output value $v = \mathbf{B}''$ of the rule base as a response to the fuzzy input value $u = \mathbf{A}'$ is calculated with $\mu_{B''}.(y) = \max_{x \in U} \min[\mu_{A'}.(y), \mu_R(x, y)]$ to $\mathbf{B}'' = \{(10; 0.1), (20; 0.3), (30; 0)\}$ according to:

$$\mu_R(x, y) : \begin{matrix} 1 & 0.3 & 0 \\ 0 & 0.3 & 0 \\ 0 & 0.6 & 1 \end{matrix}$$

$$\begin{matrix} 0.1 & 1 & 0 & \quad & 0.1 & 0.3 & 0 \\ \uparrow & & & & & \downarrow & \\ \mu_{A'}(x) & & & & & \mu_{B''}(y) & \end{matrix}$$

References

Baldwin, J. F. (1979). A new approach to approximate reasoning using fuzzy logic. *Fuzzy Sets and Systems*, 2(4), 309–325. https://doi.org/10.1016/0165-0114(79)90022-0

Bandemer, H. (1990). *Fuzzy data analysis*. Springer.

Bothe, H.-H. (1995). *Bewertung mit unscharfen Mengen*. Technische Hochschule Ilmenau.

Delgado, M. (1990). Fuzzy cardinality based evaluation of multicriteria decision making. *International Journal of Approximate Reasoning*, 4(2), 115–135.

Giloi, W. K., & Liebig, K. (1980). *Knowledge-based systems and inference structures* [German-language publication cited in context of expert systems and inference; exact source details to be verified].

Gottwald, S. (1984). *Fuzzy sets and fuzzy logic: An introduction*. Akademie-Verlag.

Mamdani, E. H. (1975). An experiment in linguistic synthesis with a fuzzy logic controller. *International Journal of Man-Machine Studies*, 7(1), 1–13. https://doi.org/10.1016/S0020-7373(75)80002-2

Mizumoto, M., & Zimmermann, H.-J. (1982). Comparison of fuzzy reasoning methods. *Fuzzy Sets and Systems*, 8(3), 253–283. https://doi.org/10.1016/0165-0114(82)90004-4

Puppe, H. (1988). *Knowledge-based systems and inference methods* [Reference cited in context of rule-based reasoning; exact publication details to be verified].

Tilli, M. (1991). *On implication operators and generalized modus ponens in fuzzy logic* [Reference details to be verified (cited in context of generalized modus ponens)].

Trillas, E. (1992). On the use of fuzzy logic in approximate reasoning. *Fuzzy Sets and Systems*, 45(1), 1–10.

Wakami, T. (1992). Fuzzy reasoning based on implication operators. *International Journal of Approximate Reasoning*, 6(2), 163–182.

Zadeh, L. A. (1973). Outline of a new approach to the analysis of complex systems and decision processes. *IEEE Transactions on Systems, Man, and Cybernetics*, SMC-3(1), 28–44. https://doi.org/10.1109/TSMC.1973.5408575.

Zimmermann, H.-J. (1991). *Fuzzy set theory – and its applications* (2nd ed.). Kluwer Academic Publishers.

Chapter 8
Fuzzy Control

Abstract This chapter extends the fuzzy rule-based expert systems with fuzzy control. This is a method of managing complex systems by using fuzzy rather than precise logic. It thereby mimics human reasoning by using linguistic variables and if-then rules to handle data that is imprecise or partially true.

The theory of fuzzy sets has long been used to solve complex control and regulation tasks. According to Bothe (1995), one of the first applications described and implemented was the control of a steam engine in laboratory operation according to Mamdani (1975). Since then, several methods with different structures have been described, but all of them are based on the rule-based expert system principle. A fuzzy controller is a very simple implementation of an rule-based expert system:

- Knowledge is represented by IF ... THEN ... rules.
- These rules are explicitly formulated by the developer for a specific, limited task and thus all relate to the current task.
- The input variables represent observations on technical systems, and the output variables are the control variables.

The fuzzy controller maps the input variables directly to the output variables on the basis of a fuzzy system description. It can therefore also be regarded as an extended engine map controller (see Preuß, 1992). The engine map represents a curved surface in a space limited by the input and output variables. The rules define individual support points of this surface, which must be completed by interpolation in such a way that even non-specified input value combinations lead to unambiguous output values. The methods of approximate reasoning presented in Sect. 7.3 can also be regarded as special interpolation methods that make it unnecessary to calculate an engine map area.

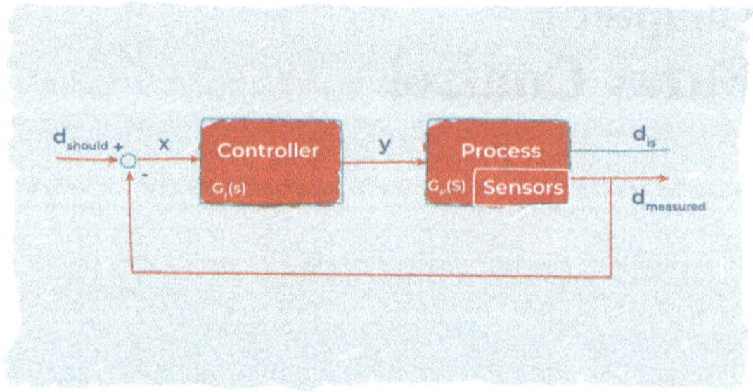

Fig. 8.1 Block diagram of a simple control loop

8.1 Guidelines for Controller Design

A basic block diagram for a simple control loop is shown in Fig. 8.1. It comprises the process to be controlled, the measuring sensors, and the controller. When a setpoint d_{set} is specified for the output value of the process, the difference $x = d_{set} - d_{meas}$ to the measured output value d_{meas} is fed back to the input of the process via the controller as a control or correction variable y. The controller must be dimensioned in such a way that it attempts to adjust the measured value to the setpoint. In some cases, a possible measurement deviation $\xi = d_{act} - d_{meas}$ should also be taken into account. A clear example of this is the idle speed control of an engine.

All variables shown in Fig. 8.1 can also be vectors. The system description can take place in both the time domain and the Laplace domain. A general design schema for simple conventional controllers in the Laplace domain can be specified as follows:

1. Determination (and simplification, such as linearization) of the transfer function $G_p(s) = d_{meas}(s)/y(s)$ of the process
2. Determination of a suitable controller transfer function $G_r(s) = y(s)/x(s)$ while complying with conditions (e.g., stability, feasibility, etc.)
3. Verification of the process and optimization of the controller

The block diagram of a fuzzy controller can be represented as shown in Fig. 8.2. Its mode of operation differs from that of a crisp engine map controller in that it uses linguistic formulations of expert knowledge about the operation of the process; the system behavior does not have to be completely known. The expert knowledge expresses how the input values y_i of the process to be controlled are to be varied for which output values x_i. Whereas in conventional map controllers the individual rules represent points in the space of input and output variables, the fuzzy rules form overlapping, evaluated areas. These are also referred to as "fuzzy points." The inference engine performs the map interpolation to determine the output variables for any premises not specified in the rules. It works on the basis of a fuzzy rule base and generates fuzzy output signals by approximate reasoning. Assuming crisp process variables and fuzzy control variables, transformations must be performed that are referred to as fuzzification and defuzzification.

Fig. 8.2 Basic structure of a
fuzzy controller

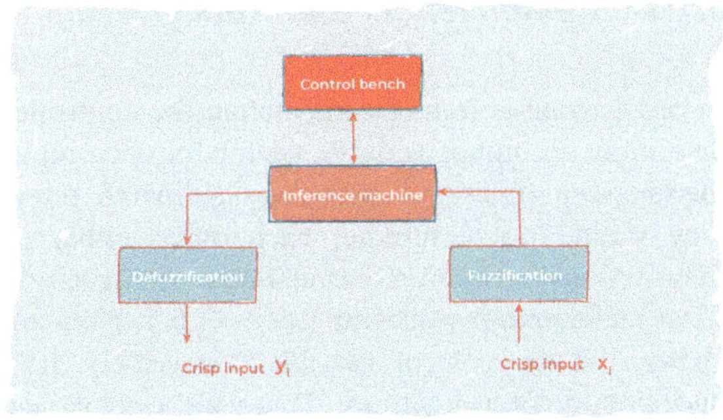

The following design schema can be used to develop a fuzzy controller:

1. Define the input and output variables:

   ↪   Which process variables should be monitored and which control variables
       should be changed?

2. Define the state interface:

   ↪   How should the linguistic variables belonging to point 1 be represented by
       terms or fuzzy sets?

3. Create the rule base:

   ↪   Which combination of rules describes the possible process states with
       sufficient accuracy?

4. Determine the methods for fuzzification and defuzzification.
5. Determine the inference method:

   ↪   According to which laws are the input values of the fuzzy controller mapped
       to the control values?

First, points 1–3 must be clarified through physical considerations and expert
interviews. This will generally be done in an iterative process that uses the results
of one point to evaluate another. Although the research to be carried out (e.g.,
questionnaires, rough system analysis, etc.) can represent a significant part of
the design work, it is very project-specific and cannot be dealt with in general
terms. Some considerations on the formulation of standardized questions and clear
linguistic variables have already been made by Zadeh first and are further explained,
for example, by Bandemer (1990) and Zimmermann (1991). They will not be
discussed in detail here. It is also possible to learn how to use the system yourself,
thereby becoming an expert and formulating the findings linguistically—and thus
not mathematically.

When defining the rule base, a few points should be noted, as described, for
example, by Pedrycz (1989) and Gariglio (1990):

8.1.1 Completeness and Rule Overlap

In order to be able to clearly define the controller behavior, the rule base should determine an output variable vector for each input variable vector. To do this, the membership functions must overlap in the relevant basic domain X. However, they should not contradict each other, either directly or in combination. The intersections between adjacent fuzzy sets, each describing adjacent terms, should lie at membership values of $\mu(x_i) = 0.5$. Non-overlapping sets can lead to abrupt changes in the control variable. Conversely, however, overlap can also lead to undesirable control actions. This can be remedied by specifying that past control actions are also used to calculate the current control actions. In this case, we refer to predictive fuzzy logic (cf. Yasunobu & Miyamoto, 1985). With predictive logic, the first rule in Example 1.6 could be rewritten as:

IF [temperature = *low*
AND cooling valve = *half open*]
THEN cooling valve = *half open*.

8.1.2 Stability

According to research by Kickert (1975), calculations in the frequency domain can be used to make statements about the stability of a controller. However, this requires the controller behavior to be severely restricted (e.g., by a linearization). Algorithms for stability analysis in general fuzzy control systems represent a significant theoretic gap, as the transfer behavior of the process to be controlled is not known or only known incompletely according to specifications. The conditions specified by Kickert therefore do not appear to be particularly practical, as there are already sufficient possible solutions for these cases in the field of "classical" control engineering. The only sure way to achieve stable controller behavior is to iteratively run through the previously presented design schema (i.e., points 1–5, with 3 only possibly).

Point 4 of the design schema (fuzzification and defuzzification) is described below, while point 5 (inference method) is highlighted due to its particular importance and explained in Sect. 8.2.

8.1.3 Fuzzification

The input values of the controller are derived from the output signals of the controlled system by sensors (e.g., strain gauges, incremental encoders, etc.). If they can be considered sufficiently accurate, they are transformed directly as in

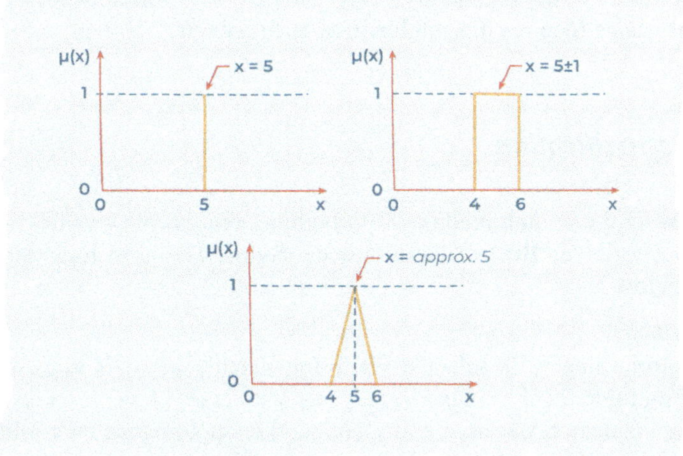

Fig. 8.3 An alternative way of defining the term "fuzzification": extension of the crisp number $x = 5$ to an interval $x' = [4, 6]$, transition to a fuzzy number $x'' = approximately$ 5 by interval evaluation

Example 1.6. This results in membership vectors whose dimension is determined by the number of terms of the corresponding linguistic variables. In this case, the inference engine further processes the membership vectors. Inaccurate (e.g., inexpensive) sensors can also cause measurement inaccuracies, which must be taken into account as evaluated error intervals. These are assigned a possibility function $\mu(x)$ based on measurement considerations (or subjective assessments). In this case, the inference engine must further process fuzzy input variables.

Example 8.1 The output voltage of a DC generator serves as a measure of the rotational speed of a loaded drive shaft. At nominal speed, the voltage $U_{nominal} = 5$ [V] is generated with a system-related error voltage of ± 1 [V]. It serves as the input signal for a fuzzy controller. If, based on previous experience but without detailed knowledge of the load torques of the drive shaft, the voltage value $x = 5$ has occurred particularly frequently, then the transition from a crisp number $x = 5$ via an interval $x' = [4, 6]$ to a fuzzy number $x'' = approximately$ 5 as a weighted interval appears plausible. The curves shown in Fig. 8.3 can also be interpreted as a probability distribution for how often the values x of the basic domain X can occur as initial values.[1]

In the case of Example 8.1, one can also speak of "fuzzification" of the crisp input value x, since the numerical value $x = 5$ is mapped to the fuzzy number or

---

[1] In order to establish a probability distribution, the process conditions must be known. Otherwise, a worst-case scenario must be considered. The following applies: what is probable must also be possible, but not vice versa.

linguistic value $x'' = approximately$ 5. The calculation of rule activation degrees for fuzzy input values requires the combination of fuzzy sets.

8.1.4 *Defuzzification*

There are several common methods for obtaining crisp control values y_0 from the fuzzy output variables $\mathbf{B}(y)$ of the inference engine. The most important ones are described below.

1. *Max method*
   The control value y_0 is selected via y for which $\mu_B(y_0)$ is maximum. In the case of multiple local maxima $y_1, \ldots, y_m$ with $\mu_B(y_1) = \ldots = \mu_B(y_m)$ or a maximum plateau between y_1 and y_m (e.g., after applying the max-min inference method), the following two methods can be used.
2. *Left (right) max method*
   The smallest (largest) y_j, $j = 1, \ldots, m$ is selected as the set value y_0.
3. *Mean value max method*
   The arithmetic mean $y_0 = 1/m \sum_{j=1,\ldots,m} y_j$ of the local maxima is selected as the control value y_0.
4. *Centroid method*
   The abscissa value of the centroid of the area below the membership function $\mu_B(y)$, $y \in Y$, is calculated as the output value y_0. Integration over Y then yields:

$$y_0 = \frac{\int_Y y\, \mu_B(y)\, dy}{\int_Y \mu_B(y)\, dy}. \tag{8.1}$$

   Since numerical integration is very complex and can lead to computing time problems, especially in real-time applications, (8.1) is usually approximated by a sum formula. One simple option is to use a weighted average of the central values $\ddot{y}_i$ of the output quantities $\mathbf{B}_i(y)$ of the individual rules (i). The activation degrees β_i of the rules (i) are used as weights. With n rules, y_0 is then calculated as:

$$y_0 \approx \frac{\sum_{i=1,\ldots,n} \beta_i\, \ddot{y}_i}{\sum_{i=1,\ldots,n} \beta_i}. \tag{8.2}$$

   The result (8.2) is exact if singletons or non-overlapping membership functions are used as consequence terms. In the centroid method, the membership functions of the two fuzzy boundary sets should be continued beyond the numerical output range, as otherwise the boundary values themselves cannot be assumed (i.e., the centroid cannot lie on a vertical boundary line of the area).
5. *Modified centroid method*

Integration in the case of 4 is only performed in the areas with $\mu_B(y) > \alpha$, $A \in [0, 1]$, $y \in Y$. Here, α represents a parameter for suppressing noise effects.

6. *Linear defuzzification*

   In the defuzzification methods presented above, all rules in the rule base theoretically contribute to the resulting fuzzy set. In contrast, in linear defuzzification, only the rule with the highest activation degree is taken into account. This simplifies the defuzzification process. For the control variable y_0, the abscissa value is then either the left or right branch of $\mathbf{B}_j$ with $\mu_{B_j}(y) = \beta_j$. The current selection depends on the specific system conditions and can, for example, take precautionary measures into account.

If there is a rule base with several rules, it is generally possible to either accumulate the partial results of the individual rules and then defuzzify them or vice versa.

8.2 Controller Behavior

8.2.1 Max-Min Inference Method

In contrast to the method described in Sect. 7.4 for solving fuzzy relational equation systems using Gödel transformation, Mamdani (1975) describes a much simpler approximate approach. According to this approach, the relational equations are first solved individually, and then a complete solution is generated from the partial solutions. In the individual solution approaches, the Gödel operation $\star$ is replaced by the Cartesian product $\otimes$ according to (7.8):

$$\mathbf{R}_i = \mathbf{A}_i \otimes \mathbf{B}_i. \tag{8.3}$$

These rule-specific individual solutions $\mathbf{R}_i$ are combined into the overall result $\mathbf{R}$ by forming a union of all rules (i.e., each rule (i) can make an equally important contribution to the result):

$$\mathbf{R} = \bigcup_i \mathbf{R}_i = \bigcup_i [\mathbf{A}_i \otimes \mathbf{B}_i]. \tag{8.4}$$

The output value $v = \mathbf{B}'' - \mathbf{A}' \circ_{\mathrm{MM}} \mathbf{R}$ of this controller for an input value $u = \mathbf{A}'$ is obtained due to the distributivity applicable according to (2.14):

$$\mathbf{A}' \circ_{\mathrm{MM}} \left[\bigcup_i [\mathbf{R}_i]\right] = \bigcup_i \left[\mathbf{A}' \circ_{\mathrm{MM}} \mathbf{R}_i\right] \tag{8.5}$$

with the aid of the max-min inference method (7.10) for individual rules to:

$$\mu_{\mathbf{B}''}(y) = \max_{x \in X} \min \left\{ \mu_{\mathbf{A}'}(x), \max_{i=1,\dots,n} \left[\min \left[\mu_{\mathbf{A}_i}(x) \right], \mu_{\mathbf{B}_i}(y) \right] \right\}$$

$$= \max_{i=1,\dots,n} \left\{ \max_{x \in X} \min \left[\min \left[\mu_{\mathbf{A}'}(x), \mu_{\mathbf{A}_i}(x) \right], \mu_{\mathbf{B}_i}(y) \right] \right\} \qquad (8.6)$$

$$= \max_{i=1,\dots,n} \left\{ \min \left[\beta_i, \mu_{\mathbf{B}_i}(y) \right] \right\}$$

With $\beta_i = \max_{x \in X} \min \left[\mu_{\mathbf{A}'}(x), \mu_{\mathbf{A}_i}(x) \right] = \text{hgt}\left[\mathbf{A}' \cap \mathbf{A} \right]$ as the activation degree of rule (i). Any crisp input values $\mathbf{A}'$ can be interpreted as singletons.

A very simple case is presented in Example 1.6 with two rules, an input variable $u = $ temperature (T) and an output variable $v = $ valve position (V). The crisp input value $T_{\text{ON}} = 18°\,$C can be interpreted as a fuzzy singleton T_{ON}. Using (8.6), $\mu_{\text{OFF}}(\varphi) = \mu_{\text{ON} \circ \mathbf{R}}(\varphi)$ is calculated as follows:

$$\mu_{\text{ON} \circ \mathbf{R}}(\varphi) = \max_{i=1,2} \left\{ \min \left[\text{hgt}(T_{\text{ON}} \cap T_i), \mu_{\varphi_i}(\varphi) \right] \right\},$$

$$= \max_{i=1,2} \left\{ \min \left[\mu_{T_i}(T_{\text{ON}}), \mu_{\varphi_i}(\varphi) \right] \right\}$$

with $T_1 = T_{\text{low}}$, $T_2 = T_{\text{medium}}$, $\varphi_1 = \varphi_{\text{half open}}$ and $\varphi_2 = \varphi_{\text{almost open}}$.

Minimum formation limits the membership functions φ_1 and φ_2 to the resulting activation degrees β_1 and β_2 of rules (1) and (2). The maximum formation for the values from both rules is performed for all relevant valve positions φ, so that the contributions of the two rules to the resulting fuzzy output value are combined.[2] The defuzzification of this fuzzy value is performed using the centroid method according to (8.1). This results in the control value $\varphi_{\text{OFF}} = 70\%$.

Note If, in the above example, the two terms of the cooling valve position are singletons with the values φ_1 and φ_2, the defuzzified output value φ_{OFF} would be calculated using the following very simple formula:

$$\varphi_{\text{OFF}} = \frac{\mu_{T_1}(T_{\text{ON}}) \cdot \varphi_1 + \mu_{T_2}(T_{\text{ON}}) \cdot \varphi_2}{\mu_{T_1}(T_{\text{ON}}) + \mu_{T_2}(T_{\text{ON}})}.$$

The following Example 8.4 shows an extended task. Here, the output variable v is influenced by two input variables u_1 and u_2 with fuzzy values. As in Example 1.6, there are now two rules.

Example 8.4 The crisp control value φ_{OFF} of the opening φ of a cooling valve is to be set according to the max-min inference method, depending on the measured crisp values T_{ON} and F_{ON} of the two input variables "temperature" and "relative

---

[2] This illustration of the max-min inference method for solving relational equation systems with the help of activation degrees of the individual rules is equivalent to the calculation using a uniform transfer relation due to (8.4).

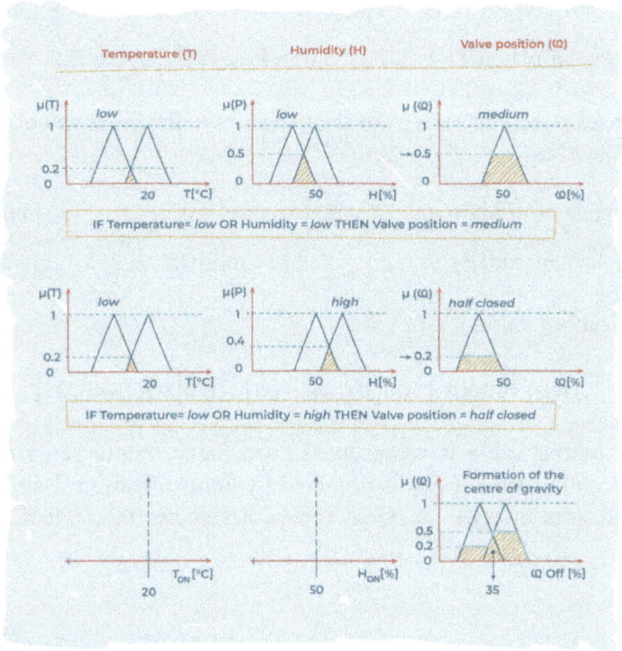

Fig. 8.4 Max-min inference method for fuzzy input variables: a cooling valve is adjusted depending on the current temperature and humidity values

humidity." For this purpose, two rules are available in linguistic form, whereby the terms used for the linguistic variables u_1 = temperature (T), u_2 = relative humidity (F), and v = valve opening (φ) are represented by fuzzy sets. These rules are as follows:

$$\text{IF T} = low \text{ OR F} = low \text{ THEN } \varphi = medium,$$
$$\text{IF T} = low \text{ AND F} = high \text{ THEN } \varphi = half\ closed.$$

A temperature $T_{ON} \approx 20[°C]$ and a relative humidity $F_{ON} \approx 50\%$ are measured and assigned rated tolerances in accordance with Fig. 8.4.

This results in the fuzzy input values $\mathbf{T}_{ON}$ and $\mathbf{F}_{ON}$. To calculate the corresponding control value ($\varphi_{OFF} \approx 20, \approx 50$), the activation degrees β_1 and β_2 of the two rules are first determined. The dependence of the valve position on T and F is taken into account by forming the maximum for the resulting activation degree β_1 of the first rule (OR operation within the condition part) and by forming the minimum for the resulting activation degree β_2 of the second rule (AND operation within the premise). This results in:

$$\beta_1 = \max\big[\mathrm{hgt}(\mathbf{T}_{\mathrm{ON}} \cap \mathbf{T}_{\mathrm{low}}),\ \mathrm{hgt}(\mathbf{F}_{\mathrm{ON}} \cap \mathbf{F}_{\mathrm{low}})\big] = 0.5 \quad \text{and}$$

$$\beta_2 = \min\big[\mathrm{hgt}(\mathbf{T}_{\mathrm{ON}} \cap \mathbf{T}_{\mathrm{low}}),\ \mathrm{hgt}(\mathbf{F}_{\mathrm{ON}} \cap \mathbf{F}_{\mathrm{high}})\big] = 0.2.$$

The responses $\mu_1(\varphi)$ and $\mu_2(\varphi)$ of the two rules to the input values $\mathbf{T}_{\mathrm{ON}}$ and $\mathbf{F}_{\mathrm{ON}}$ are thus calculated as:

$$\mu_1(\varphi) = \min\big[\beta_1, \mu_{\varphi_{\mathrm{medium}}}(\varphi)\big] = \min\big[0.5, \mu_{\varphi_{\mathrm{medium}}}(\varphi)\big] \text{ and}$$

$$\mu_2(\varphi) = \min\big[\beta_2, \mu_{\varphi_{\mathrm{half\ closed}}}(\varphi)\big] = \min\big[0.2, \mu_{\varphi_{\mathrm{half\ closed}}}(\varphi)\big],$$

and the fuzzy output value $\mu_{12}(\varphi)$ as:

$$\mu_{12}(\varphi) = \max\big\{\min\big[\beta_1, \mu_{\varphi_1}(\varphi)\big],\ \min\big[\beta_2, \mu_{\varphi_2}(\varphi)\big]\big\}.$$

The fuzzy output value is represented by a fuzzy output set, from which the required crisp output value φ_{OFF} is obtained by defuzzification. Using the centroid method, this results in $\varphi_{\mathrm{OFF}} = 35\%$. Figure 8.4 shows this calculation rule as a diagram.

8.2.2   Max-Prod Inference Method

Another way to simplify the Gödel inference method according to Sect. 7.4 was pursued by Holmblad and Østergaard (1982) in the control of a cement kiln, where they used a multiplication in (8.4) to calculate the Cartesian product. The output value $v = \mathbf{B}''$ of the controller is then given by:

$$\mathbf{B}'' = \bigcup_i [\beta_i \cdot \mathbf{B}_i] = \bigcup_i \big[\big[\mathrm{hgt}(\mathbf{A}' \cap \mathbf{A}_i)\big] \cdot \mathbf{B}_i\big], \tag{8.7a}$$

with the membership function

$$\mu_{\mathbf{B}''}(y) = \max_{i=1,\dots,n} \big\{\beta_i \cdot \mu_{B_i}(y)\big\}. \tag{8.7b}$$

This procedure describes the max-prod inference method for solving relational equation systems.

While the methods described work according to the principle of "accumulation after fuzzification," the partial results in the following two heuristic methods are first fuzzified and then accumulated. This approach is motivated by the fact that this is the only way to ensure that the partial results are weighted according to their frequency of occurrence.

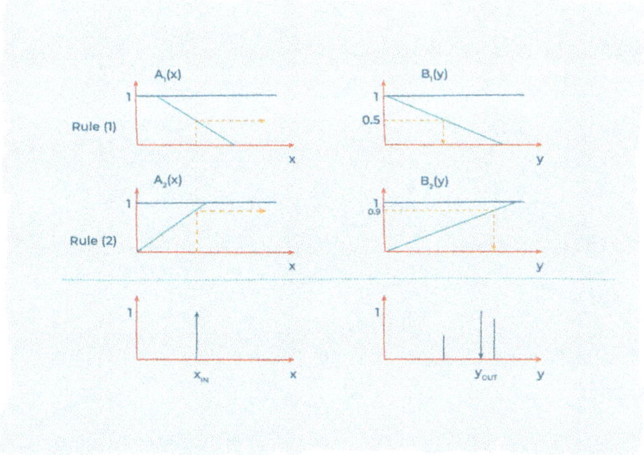

Fig. 8.5 Tsukamoto inference method with two rules and one input variable

8.2.3 Tsukamoto Inference Method

This method requires the use of monotonic membership functions for the control variables (as shown in Fig. 8.5). This directly produces numerical partial results that do not need to be fuzzified.

The activation degrees β_i of the individual rules are calculated as:

$$\beta_i = \mu_{\mathbf{A}_i}(x_{\mathrm{IN}}). \tag{8.8}$$

The numerical partial results for the individual rules are obtained by applying the inverse mapping $\mu_{\mathbf{A}_i}^{-1}$ of the membership function to:

$$y_i = \mu_{\mathbf{A}_i}^{-1}(\beta_i). \tag{8.9}$$

The overall result is then calculated by weighted averaging to:

$$y_i = \frac{\sum_i (\beta_i y_i)}{\sum_i (\beta_i)}. \tag{8.10}$$

8.2.4 Sugeno Inference Method

The Sugeno inference method is illustrated for two rules and one input variable in Fig. 8.6.

In contrast to the rules discussed so far, the Sugeno-type rules read as follows:

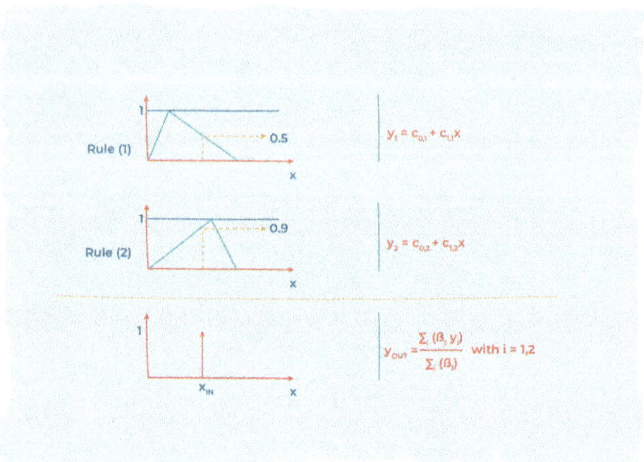

Fig. 8.6 Sugeno inference method with two rules and one input variable

$$\text{IF } u = \mathbf{A}_i \ \text{ THEN } y_i = s_{0,i} + s_{1,i}\, x_{\text{IN}}. \tag{8.11}$$

The output therefore directly produces numerical values that represent the linear combination of the input values (in the above example, only x_{IN}). The activation degrees β_i of the rules (i) are calculated by fuzzifying the input values to:

$$\beta_i = \mu_{\mathbf{A}_i}(x_0). \tag{8.12}$$

The partial results of the individual rules are already specified in the consequence part with:

$$y_i = \sum_k \left(s_{k,i}\, x_k \right), \tag{8.13}$$

where the index k is equal to the number of input variables to be considered in the condition part (i.e., $k = 1$ in the example shown in Fig. 8.16). As with the Tsukamoto inference method, the overall result is obtained by weighted averaging to:

$$y_{\text{OUT}} = \frac{\sum_i (\beta_i y_i)}{\sum_i (\beta_i)}. \tag{8.14}$$

It should be noted that this result is already available numerically and does not need to be fuzzified in order to obtain a crisp control variable. The degrees of freedom $s_{k,i}$ must be adjusted. Optimization can be carried out very effectively, for example, with the help of genetic algorithms or in a learning process using artificial neural networks (Bothe, 1995).

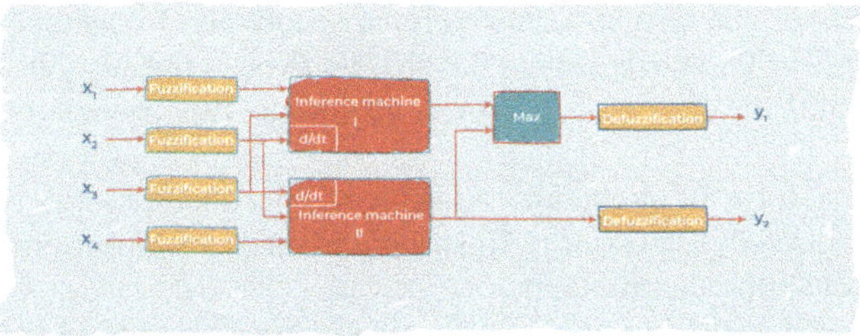

Fig. 8.7 Example block diagram of a fuzzy PD controller with four measured variables and two control variables based on two inference engines

If only the constant parameters $s_{0,i}$ are not equal to 0 (i.e., $y_i = s_{0,i}$ applies), the same output result is obtained as with the max-min or max-prod inference method if only singletons $s_{0,i}$ appear in the rules as consequence terms. A more in-depth discussion can be found, for example, in Buckley (1993).

When constructing the rule base, it may also happen that contradictions arise in certain process situations after combining some production rules. If individual rules are to be retained in order to describe other states, the aim must be to minimize the resulting contradictions for particularly important statements or states. For this purpose, the rules can, for example, be organized into groups, and separate fuzzy relations can be calculated for these groups. This creates a system of several fuzzy inference engines working in parallel. The following section presents a basic example of this case.

Example 8.5 A proportional-differential (PD) controller[3] is to be designed that converts the four measured variables x_1, x_2, x_3, and x_4 into two control variables y_1 and y_2. The input variables are first fuzzified by forming vectors of sympathy. Further processing can be carried out as shown in Fig. 8.7 by dividing the rule base into two inference engines operating in parallel. The first derivatives of the input variables x_1 and x_2 are also taken into account when calculating the control variables. The inference engines deliver fuzzy intermediate results, which are combined for y_1 using a maximum operator. After defuzzification, the crisp control variables (1) and y_2 are obtained; the entire control set applies to y_1, while only the second control set applies to y_2.

The rules of the two inference engines must be defined in such a way that the map in the space of input and control variables results in an area that is similar to that of conventional systems. In contrast, however, this results in a larger number of adjustable degrees of freedom, so that the theoretically conceivable optimal

---

[3] The transfer behavior of a PD controller includes both a proportional element, which is determined by a gain factor, and a differential element, which is determined by a time constant. The design of a fuzzy PD controller is discussed in the last application example and in Driankov et al. (1993), Hayashi (1991), He et al. (1993), Lee (1992), and Malki et al. (1994).

controller behavior can be approximated more accurately. Depending on the extent and accuracy of the knowledge about the process to be controlled, the map is varied interactively until the desired behavior is achieved. This process is conveniently carried out on a computer-aided development system. If process knowledge is available in the form of probability statements, the membership functions used should be interpreted as possibility distributions and, according to Theorem 5.2, include the probability distributions.

Simulation can also be used in conjunction with a variation of the membership functions to construct model rule bases that attempt to replicate specified expert statements. Once the desired accuracy has been achieved, the fuzzy controller is implemented either in hardware or in software.

8.3 Application Examples

8.3.1 Inverted Pendulum

A large number of authors have addressed the problem of the inverted pendulum. One of the first solutions comes from Cannon (1966). The problem was extended to double pendulums (i.e., Sturgeon & Loscutoff, 1972) and triple pendulums (Furuta et al., 1984). It is also a frequent topic in control engineering training for engineers (e.g., Hartmann, 1992).

The task is representative of a series of unstable processes that are to assume a desired behavior through the application of control engineering methods. The controlled system consists of a cart on which a pole is to be balanced; the pole is mounted with virtually no friction and is attached (as shown in Fig. 8.8). The cart is moved by a motor.

The pole can only move in the plane around the pivot point, which is defined by the vertical axis and the direction of movement of the cart. The cart and pole system behaves unstably in the arrangement shown in Fig. 8.8: A small angular deflection $e(t)$ caused by gravity g gives rise to an angular acceleration $e''(t)$ of the pole, which in turn increases $e(t)$. The task of the motor control is now to balance the pole by moving the cart. To do this, the angular displacement $e(t)$ of the pole and the position $x(t)$ of the cart are measured; by differentiation, the controller can also be provided with the angular velocity $e'(t)$, the velocity $x'(t)$, and the acceleration $x''(t)$. The controller output controls the motor.

A specific position x_{target} of the cart is targeted, as otherwise balancing would also be possible at a constant speed and the cart would violate the actual spatial limitations.

First, the process of controller design using classical methods will be demonstrated; the actual design will only take place after a system analysis of the controlled system (i.e., the dynamic behavior of the cart and pole system). There are

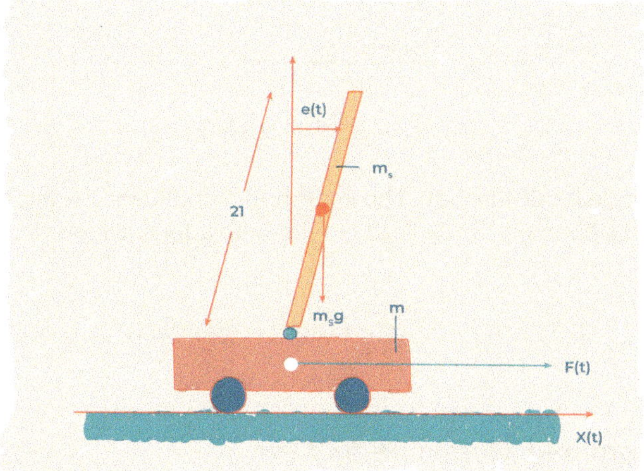

Fig. 8.8   Test setup for stabilizing an inverted pendulum

a variety of different design methods to choose from, but these will not be described in detail here.

8.3.2   Classic Control

First, the equations of motion for the pole and for the cart are established separately, resulting in a system of two differential equations. In the linear case, the transfer functions of the system can be calculated in the Laplace domain, on the basis of which the desired controller can be designed. In the following, we will limit ourselves to establishing the equations of motion. For controller design, please refer to the usual textbooks on control engineering, such as Landgraf (1970). For simplicity, the power transmission from the motor to the cart is assumed to be linear.

8.3.3   Equation of Motion of the Pole

According to Steiner's theorem, the moment of inertia Θ_s of the pole around the pivot point:

$$\Theta = 4/3 \, m_s \, l^2, \tag{8.15}$$

where m_s is the total mass of the pole and l is half the pole length. The gravitational force $g = 9.81$ m/s^2 causes a torque D_M at the pivot point when the pole is deflected, with:

$$D_M = l\, m_s\, g \sin e(t). \tag{8.16}$$

The acceleration of the cart, on the other hand, acts with the torque:

$$D_B = -l\, m\, x''(t) \cos(e(t)), \tag{8.17}$$

where m is the mass of the cart. The equilibrium condition for the torques at the pivot point, together with (8.15), leads to the following equation of motion for the pole:

$$4/3\, l\, e''(t) + x''(t) \cos(e(t)) - g \sin(e(t)) = 0. \tag{8.18}$$

8.3.4 *Equation of Motion of the Cart*

The following horizontal forces act on the cart of mass m:

—Tangential force of tilting pole: $\qquad F_t = -m\, l\, e''(t) \cos e(t), \qquad (8.19)$

—Radial force of tilting pole: $\qquad F_r = m\, l\, e'(t)^2 \sin e(t), \qquad (8.20)$

—Friction force: $\qquad -F_\mu = -c_\mu\, x'(t), \qquad (8.21)$

—Applied motor force: $\qquad F(t). \qquad (8.22)$

The factor c_μ in (8.21) represents a friction constant. Together with the acceleration force $(m + m_s)\, x''(t)$, this results in the force weight:

$$(m + m_s)\, x''(t) = -c_\mu x'(t) + F(t) - m_s l\, e''(t) \cos(e(t)) +$$
$$+ m_s l\, e'(t)^2 \sin(e(t)). \tag{8.23}$$

The task of the motor control is therefore to balance the pole by varying the drive force $F(t)$ of the motor so that $e(t) \approx 0$. With (8.18) and (8.23), the cart and pole system to be controlled is highly nonlinear at larger pole deflections. However, it can be linearized around the rest position (i.e., for the conditions $|e(t)| \ll 1$ and $m_s \ll m$) to:

$$4/3\, l\, e''(t) + x''(t) - g\, e(t) = 0 \tag{8.24}$$

$$m\, x''(t) = -c_\mu x'(t) + F(t). \tag{8.25}$$

For these linear differential equations, a controller can be designed using the root locus method, for example. However, for larger pole deflections, linearization according to (8.24) and (8.25) is not permissible, since larger areas of the nonlinear

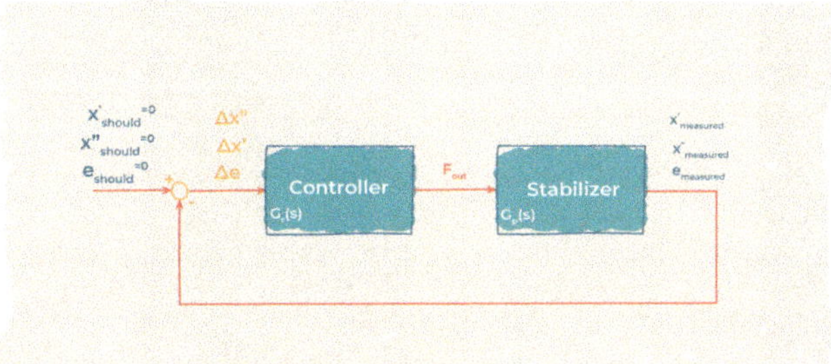

Fig. 8.9 Control loop for the cart and pole system

characteristic curve of the controlled system must be traversed for stabilization; the more nonlinear the real behavior of the controlled system becomes, the more likely it is that undesirable boundary oscillations will occur in the controller. In this case, much more complex algorithms must be used that are also suitable for controlling nonlinear systems. The basic block diagram for the control loop is shown in Fig. 8.9.

When deriving (8.12) and (8.17), a number of idealizing assumptions were made about the cart and pole system; these mean that the controller must always be adapted to the actual system behavior through an optimization process. The idealizations include, for example, the following additional assumptions:

- No friction at the pivot point of the pole
- Negligible air friction
- Rigidity and homogeneous mass distribution of the pole
- Force transmission to the ground through static friction
- Negligible pole mass compared to the cart mass
- Small deflection angle from the position to be balanced
- Small measurement errors in displacement and angle values

The facts described can also be interpreted in terms of the definition of the terms "system" and "modeling " in Sect. 1.1 in such a way that certain coupling variables of the cart and pole system with the environment are not recorded in the system analysis. The division of the system into subsystems is used to calculate its motion behavior and is based on significant simplifications. The subsystems themselves are also described in a highly simplified manner. Since a similar approach is also typical for controller design using classical methods with a more refined representation of reality, it cannot be said that this is a "rigorous" approach. The following quote by Einstein (1921) is representative of these "real-world problems": "As far as the laws of mathematics refer to reality, they are not certain; and as far as they are certain, they do not refer to reality."

IF			THEN	
E = NM	and	D = NM		
ΔX = NM			THEN	K = NG
E = NG				
E = NM	and	D = NK		
E = NK	and	D = NM		K = NM
E = ZR	and	D = NM		
ΔX = NK				K = NK
E = NK	and	D = NK		
E = PM	and	D = NM		
E = ZR	and	D = ZR		K = ZR
E = NM	and	D = PM		
ΔX = PK				K = PK
E = PK	and	D = PK		
E = PM	and	D = PK		
E = PK	and	D = PM		K = PM
E = ZR	and	D = PM		
E = PM	and	D = PM		
ΔX = PM				K = PG
E = PG				

Fig. 8.10 Rule base for stabilizing a cart and pole system

8.3.5 *Fuzzy Control of the Cart and Pole*

In contrast to the design process described above, fuzzy control does not require quantitative knowledge of the dynamic system behavior; the description is qualitative, based on linguistic expressions. The fuzzy logic approach takes advantage of the fact that, after a relatively short period of practice, humans are able to balance a pole on their hand without having a quantitative understanding of the underlying system equations. They unconsciously use rules such as "If the pole tilts to the right, I move my hand slightly to the right."

To establish a rule base for the inverted pendulum, the four linguistic variables "angle error" E, "rotational speed" D, "position error" ΔX, and "force" K are to be introduced. Simulation results from Brinkmann and Moraga (1992) suggest that the cart speed can initially be considered subordinate. E and K are each represented by seven terms, D and ΔX by five terms. The following abbreviations are used for this in the rules listed:

$$PG = \textit{positive large}, NK = \textit{negative small},$$
$$PM = \textit{positive medium}, NM = \textit{negative medium},$$
$$PK = \textit{positive small}, NG = \textit{negative large},$$
$$ZR = \textit{normal}.$$

According to the proposal by Brinkmann and Moraga (1992), a rule base with 19 individual rules is set up, which are shown in Fig. 8.10.

E \ D	NM	NK	ZR	PK	PM
NG	NG	NG	NG	NG	NG
NM	NG	NM			ZR
NK	NM	NK			
ZR	NM		ZR		PM
PK				PK	PM
PM	ZR			PM	PG
PG	PG	PG	PG	PG	PG

ΔX	NM	NK	ZR	PK	PM
	NG	NK		PK	PG

Fig. 8.11 Matrix form of the rule base of the cart and pole system

Rules 2, 7, 12, and 18, which relate to the position x, have a more indirect effect: In order to move toward the target position x_{target}, the cart must first be accelerated briefly in the opposite direction so that the pendulum tilts to the necessary angle. When x_{target} is exceeded, the cart is accelerated further, the pendulum tilts in the opposite direction, and the cart can move backward. Since the problem is symmetrical with respect to the target position, the rule base is also symmetrical.

The rule base can also be represented in matrix notation, where the columns and rows are determined by the terms of the measured variables, and the terms of the control variable appear as possible matrix elements. This leads to a representation of the rule base as shown in Fig. 8.11.

In the following, the membership functions of the fuzzy sets belonging to the terms will be presented. For the linguistic variables "angle error" and "force," the resolution near the zero point is particularly high in order to be able to react early and in a measured manner to a slight tilt of the pole (Figs. 8.12, 8.13, 8.14, and 8.15).

8.3.6 Control of a Cement Kiln

This chapter will explain in more detail the example of fuzzy control of a cement kiln mentioned in Sect. 8.2. Cement is produced by heating a mixture of clay, ground sandstone, sand, and iron ore until the compounds C_2S, C_3S, C_3Al, and $C_4Al\,F$

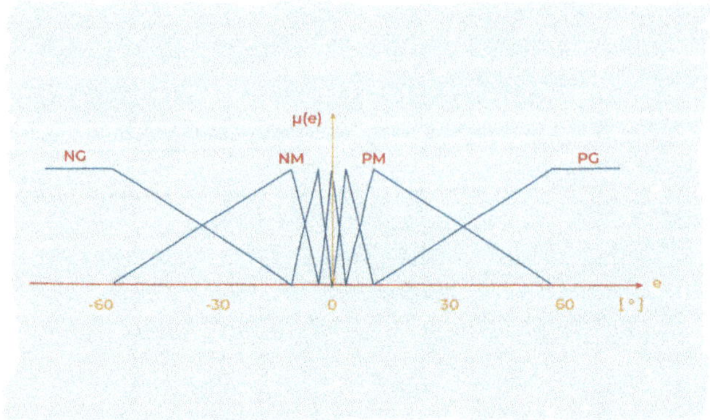

Fig. 8.12 Terms of the linguistic variable "error angle" E

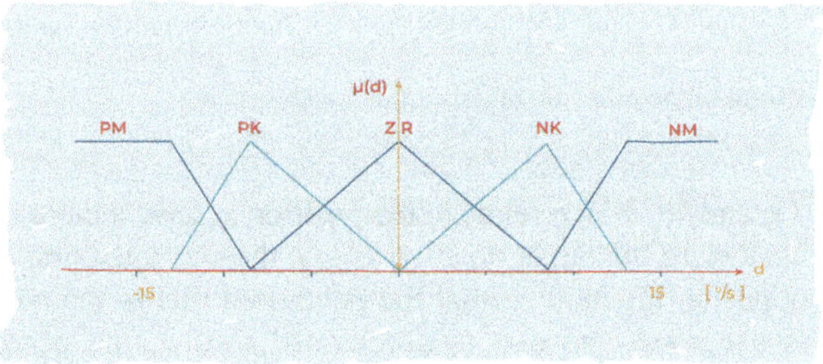

Fig. 8.13 Terms of the linguistic variable "rotational speed" D

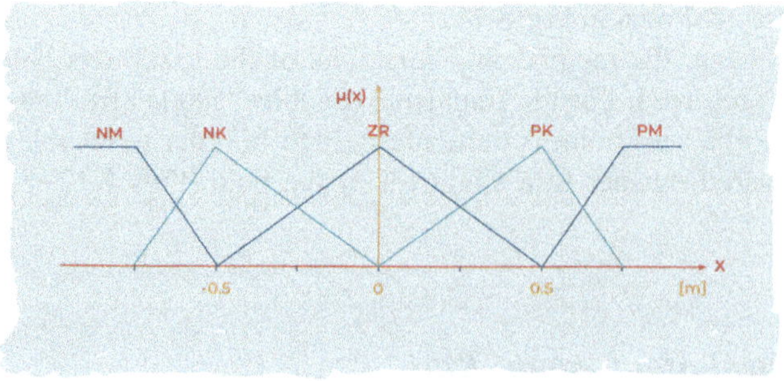

Fig. 8.14 Terms of the linguistic variable "error position" ΔX

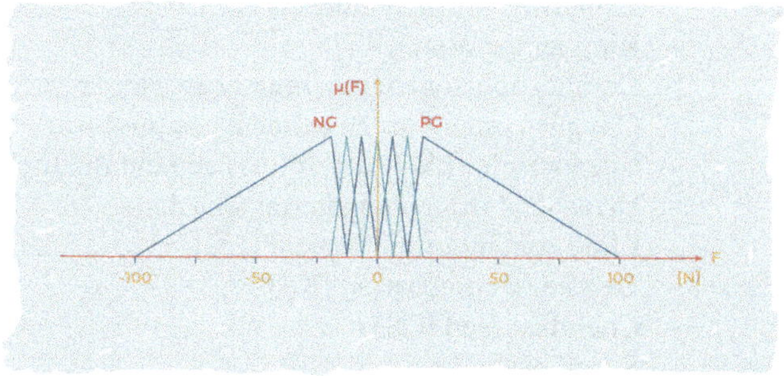

Fig. 8.15 Terms of the linguistic variable "force" K

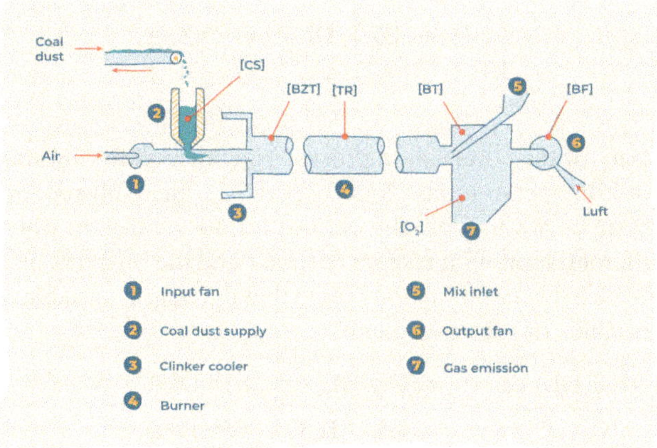

Fig. 8.16 Simplified circuit diagram of a cement kiln (according to Zimmermann, 1991)

are formed. The burning process consists of three main stages: drying process, calcium carbonate decomposition, and actual burning process at 1250–1450°C. Figure 8.16 shows a simplified cross-sectional view of the kiln. First, let us explain the equipment in more detail:

Firing tube: – Steel sleeve approx. 130 m long and 5 m in diameter
                  – Suspended at a slight angle
                  – Rotates at approximately 1 rpm
Material speed: – 3 hours 15 minutes per 130 m length
                  – 45 minutes dwell time in the clinker cooler
Heating: – Combustion of a coal dust-air mixture
                  – Propagation direction opposite to the material movement

Problems arise with automatic control using conventional methods because the system contains time delays and behaves in a time-variant and nonlinear manner. The most important input and output variables of the controller are:

Input variables: – Exhaust gas temperature [BT]
                        – Ring temperature [RT]
                        – Combustion zone temperature [BZT]
                        – Oxygen content of the exhaust gas [O_2]
                        – Liter weight [LW] (determines cement quality)
                        – Drive load (moving material) [DL]
                        – Lime content of the cement [CC]
Output variables: – Degree of combustion [KS]
                        – Coal dust feed [CS]
                        – Speed of the final fan [BF]

Excerpt from the existing process knowledge (increase: ↑, decrease: ↓):

$$- \text{CS} \uparrow \Rightarrow \text{BZT, DL} \uparrow; O_2, \text{CC} \downarrow.$$
$$- \text{BF} \uparrow \Rightarrow \text{BZT, CC} \uparrow; O_2, \text{DL} \downarrow.$$

$$\cdots$$

The complexity of the combustion process is evident. After thorough discussion with the cement kiln operators (as "local experts" in their sub-process), 75 rules of the following basic form were initially defined for the controller, where the operator d/dt denotes differentiation with respect to time t and the brackets on the right-hand side indicate the possible combinations (note that the output variables of the process are the input variables of the controller):

IF d(DL)/dt   = ⟨DL, SL, OK, SH, DH⟩

AND DL   = ⟨DL, SL, OK, SH, DH⟩

AND BZT   = ⟨L, OK, H⟩

THEN d(O_2)/dt = ⟨VN, N, SN, ZN, OK, ZP, SP, P, VP⟩

PLUS d(BF)/dT = ⟨VN, N, SN, ZN, OK, ZP, SP, P, VP⟩.

The different value quantization in five, three, and nine sequences can be seen. For the input variables u, seven terms are given with the abbreviations:

DL = very small,   SH = fairly large,
L = small,   H = large,
SL = fairly small,   DH = very large,
OK = normal.

For the output variables v, nine terms are given with the abbreviations:

VN = very negative   ZP = slightly positive
N = negative   SP = fairly positive
SN = fairly negative   P = positive
ZN = slightly negative   VP = very positive
OK = no need for regulation

These linguistic characteristics are represented by membership functions $\mu_i(x_i)$ with four discrete levels in the value range [0,1] and 15 discrete levels in the base set X_i. The inference method used (and developed for this task) was described in Sect. 8.2. The number of rules was then reduced in an optimization process.

8.3.7 Use of Fuzzy Methods in a Single-Lens Reflex Camera

Many photographers require their camera to deliver successful images in every possible environment. This should apply to spontaneous snapshots, fast-moving objects, and all conceivable photographic situations. Complex microprocessor controls are used for the tasks involved, especially in the automatic programs of high-quality single-lens reflex cameras. Fuzzy methods, which use exclusive expert knowledge, are increasingly being used in this consumer goods sector in particular; they can lead to faster response times for the camera electronics with lower hardware costs.

The following section describes the control concept of a high-quality single-lens reflex camera, which uses fuzzy logic methods for essential automatic control processes. According to the implementation notes in Akahoshi (1991) and Norita (1992), this can significantly reduce the number of control rules that need to be used for precise situation analysis and determination of control values, compared to a crisp rule-based expert system. In a varying scenario (i.e., changing sensor readings), a smooth control behavior is automatically created. Specifically, fuzzy methods are used in the following four areas:

- Autofocus setting (AF):   determines the distance of the main object within the image.
- Exposure control (AE):   determines the exposure factor, taking into account the main object and background.
- Expert program (AP):   determines the optimal values for aperture and exposure time, taking into account the exposure factor.
- Zoom control (AZ):   controls the zoom speed via a built-in motor.

During the development phase, a photographic knowledge base was first created, and the behavior of the individual fuzzy controllers was optimized. The knowledge base contains prototypical model situations (with the main objects in different positions, sizes, etc.) against different background patterns. Professional photographers determined the typical settings for these situations and stored them together with the corresponding measurements from the camera sensors. The behavior of the fuzzy controllers was then simulated using an interactive fuzzy development system. In an iterative process, the membership functions and rules were varied until the controllers delivered the desired expert settings.

The finished overall system was implemented on a microcontroller (as shown in Fig. 8.17). The individual controllers were combined into a uniform module that handles all tasks.

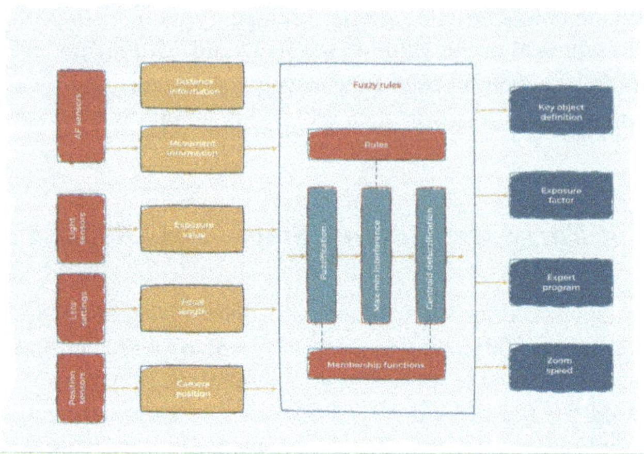

Fig. 8.17 Block diagram of the fuzzy control of a single-lens reflex camera

The fuzzy controller uses the max-min inference method and the centroid method for defuzzification. The inference engine of the overall system occupies approximately 500 bytes of memory in the program area. The input variables are interpreted as linguistic variables and represented with up to five terms. The membership functions are stored as tables with 16 individual values in 256 gradations.

The rules are designed as structures that refer to these tables with pointers; the microcontroller calculates the conclusions by tracing these structures. Each block of two rules and two inputs requires a data area of approximately 90 bytes.

The input variables for the fuzzy controller are the object distance information and the motion information from the autofocus sensor system, brightness information measured in 14 areas, the currently set focal length, and the maximum aperture[4] of the lens, as well as orientation information regarding the horizontal and vertical camera position.

The four individual applications are described in detail below, including the sensor technology used to record the measured values.

8.3.8 Autofocus Setting

The distance of the object to be focused on from the film plane is determined using an approximate conclusion. In the center of the viewfinder image, four distance sensors form a narrow and a wide focus frame (as shown in Fig. 8.18a).

If the individual sensors detect different distances, fuzzy logic determines which sensor represents the current value. To do this, truth values for the existence of the

---

[4] The maximum aperture of a lens is the aperture value with the largest possible aperture opening.

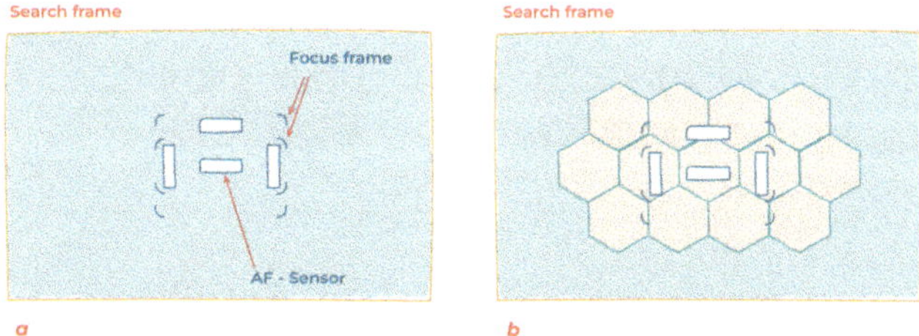

Fig. 8.18 (**a**) Arrangement of autofocus sensors. (**b**) Exposure measurement with 14-segment brightness cells

main object in the range of each sensor are first calculated for each autofocus sensor using empirical rules, membership functions, and information from the position sensors. The maximum truth value indicates the position of the main object in the sensor field, and the active sensor determines its distance from the film plane and thus the necessary distance setting.

8.3.9   Exposure Control

A silicon photocell with spectral sensitivity similar to that of the human eye is used as the sensor for exposure measurement. As shown in Fig. 8.18b, it is divided into 13 honeycomb cells in front of an integrating background area and provides the brightness values at the corresponding points as measured values.

In the "exposure control" module, an exposure factor is calculated that correlates with the current film sensitivity. This is done as follows (as shown in Fig. 8.19): Based on the measured values from the exposure sensors and the position of the main object, the "FUZZY-1" and "FUZZY-2" submodules determine a background factor and a main object factor. "FUZZY-3" submodule calculates the exposure factor to be set from this information and the original measured values.

This is forwarded to the "expert program" module, which uses it to calculate a combination of aperture value and exposure time that is particularly favorable for the photographic scene. A large aperture value represents a small opening of the lens aperture.

8.3.10   Expert Program

The basic relationship between exposure factor, aperture value, and exposure time is shown in Fig. 8.20. The exposure time is plotted on the x-axis in a logarithmic representation, decreasing toward larger x-values, while the aperture value is plotted on the y-axis. The diagonal lines indicate the lines of constant exposure factors. If

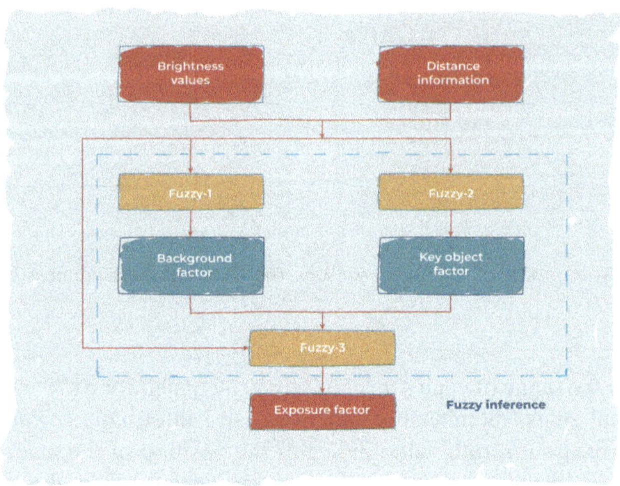

Fig. 8.19 Calculation of the exposure factor

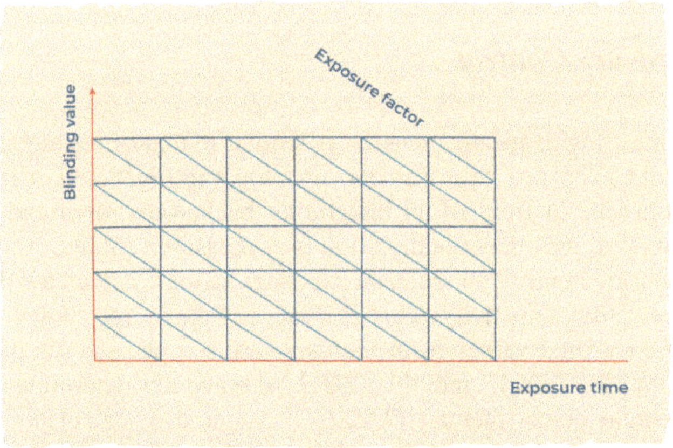

Fig. 8.20 Exposure diagram for exposure time and aperture value

a specific exposure factor is specified, setting the aperture value or exposure time determines the other variable.

While the exposure time is specified in aperture priority mode, the aperture must be preselected in the shutter priority mode. In addition to these methods, an expert program has been implemented in the camera that determines a particularly favorable combination of exposure time and focal length based on the image scene data, the lens and camera data, and the stored expert knowledge. A block diagram for the calculation algorithm is shown in Fig. 8.21.

First, the inference engine determines the membership values of the current image to the terms of a linguistic variable "scene type." For this purpose, a variable

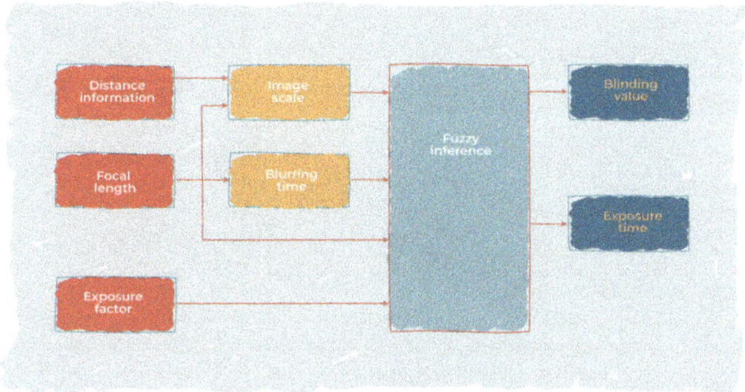

Fig. 8.21 Calculation of exposure time and aperture value

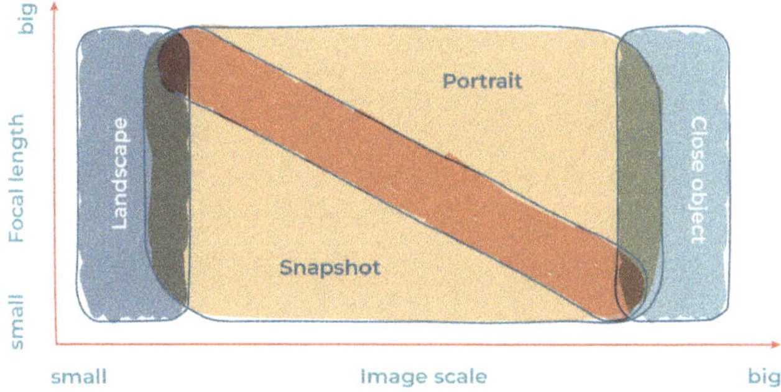

Fig. 8.22 Formation of the linguistic variable "scene type"

"image scale" is introduced, which estimates the real object size based on the current focal length and distance information. In addition to this image scale, the inference engine also takes into account the focal length-dependent exposure time, which compensates for camera shake caused by a shaky hand, as well as the exposure factor. The exposure time and aperture value are specified directly at the output. The ranges for the four scene types "landscape," "portrait," "snapshot," and "close-up" are shown in Fig. 8.22.

The rules of the inference engine also relate to the desired depth of the field of the photo, which can be derived from the scene type and determines the aperture value when the current focal length is specified. Figure 8.23 shows the basic tendencies according to which the rules specify the aperture value to be set.

Using the linguistic variables "focal length" f, "image scale" m, and "aperture value" b, rules of the form:

$$\text{IF } f = small \text{ AND } m = medium\ small \text{ THEN } b = semi\text{-}closed.$$

For b, the linguistic terms *closed*, *semi-closed*, *medium*, *semi-open*, and *open* are available. If the exposure factor is set to EV= 14, the exposure situation shown in Fig. 8.24 results. With the lens used in this example, numerical aperture values in

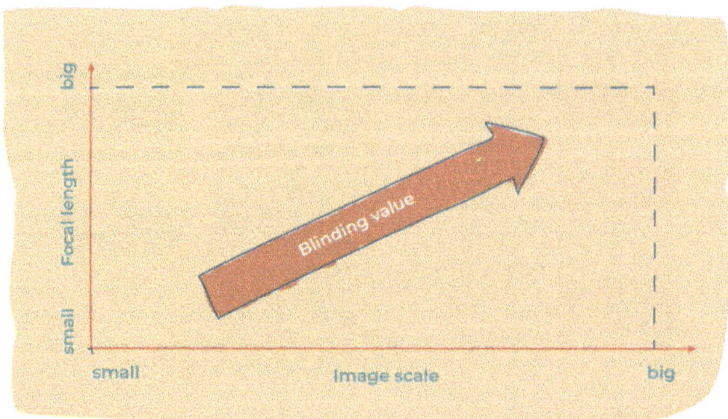

Fig. 8.23 Structure of the rule base for setting the aperture value with regard to a depth of the field that is appropriate for the scene

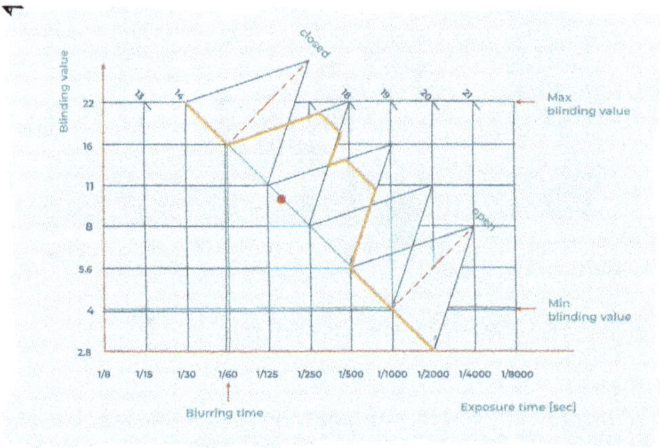

Fig. 8.24 Calculation of exposure time and aperture value using the centroid method. The terms are specified above the two outer membership functions

the range $4 \leq b \leq 16$ can be set, and the current focal length results in a time during which an image could be blurred of 1/60 sec. The shortest exposure time of the camera is 1/8000 sec. This defines the framework conditions for possible settings of the expert program. The terms must now be mapped to the basic set of possible settings using fuzzy sets in such a way that exposure times between 1/60 sec and 1/1000 sec and thus aperture values between 16 and 4 are possible.

The membership functions of the terms for the linguistic variable "aperture value" are also plotted in the exposure diagram. The peak values for the two outer terms are determined by the value combinations (1/60 sec, 16) and (1/1000 sec, 4); because they are both of the triangle type, these points determine the adjustable limit values in defuzzification according to the centroid method.

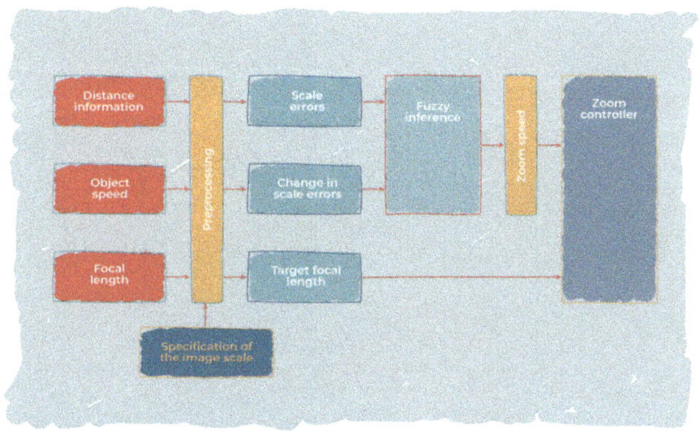

Fig. 8.25 Fuzzy zoom control

With a result of the rule from evaluation according to the bold membership function, the marked point is calculated as a particularly favorable setting value. In this example, it is an exposure time of 1/150 sec and an aperture value of 10.

8.3.11 Zoom Control

The camera has an "image size lock mode" for automatic zoom adjustment when photographing moving objects. In this setting, the focal length of the zoom lens is continuously changed so that the image scale (i.e., the size of the selected image section) is maintained. A block diagram of the corresponding zoom control is shown in Fig. 8.25.

The input variables are the object distance information, the corresponding temporal distance changes, the focal length, and the desired set image scale. Signal preprocessing is used to calculate the variables "scale error," "temporal change in scale error," and "target focal length." The target focal length is a value calculated by prediction for the distance of the main object.

The scale error and its temporal change serve as input variables for an inference engine, which delivers the zoom speed to be set as the output variable. Together with the target focal length, its value determines the motor control for focal length adjustment. The basic concept of the rules for determining the zoom speed is shown in Fig. 8.26.

The zoom speed is accelerated for the lower-left and upper-right corners of the diagram. An example production rule can be specified as follows using the scale error r and the focal length change $\partial f/\partial t$:

IF $r = zero$ AND $\partial r/\partial t = slightly\ increasing$ THEN $\partial f/\partial t = negative\ low$.

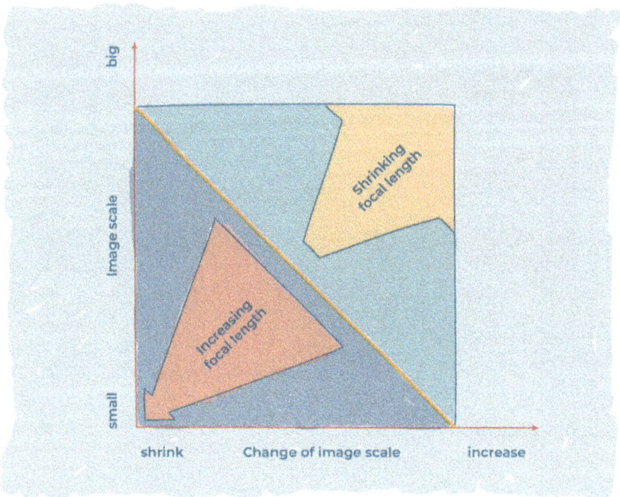

Fig. 8.26 Control concept for calculating the zoom speed

8.3.12 Fuzzy Proportional Integral Derivative Controller

The majority of automatic controllers used in the industry today are conventional three-term or proportional integral derivative (PID) controllers. They are used in process engineering plants or for temperature control, as well as in the automotive sector or for controlling electrical machines. Similar to the definition of a PD controller, the transfer behavior of a PID controller consists of the superposition of a proportional, an integral, and a differential component. A block diagram is shown in Fig. 8.27.

The proportional signal component of the manipulated variable is generated by multiplying the input signal by a constant C_p, the integral component by integrating over time and then multiplying by a constant $C_i/\Delta t$, and the differential component by differentiating with respect to time and then multiplying by a constant $C_D \Delta t$. In the case of a time-discrete controller, integration is replaced by a (finite) summation and differentiation by the formation of the difference quotient. The output signal of the integrator must also be limited in order to prevent the value range from being exceeded. This is indicated in Fig. 8.27 by the series connection of a block with a sigmoidal transfer function.

All three branches can be simulated using fuzzy inference machines. However, this results in a very high computational effort, as there are then a total of three inputs and the possible number of rules increases exponentially. If, for example, the value ranges are partitioned with seven linguistic terms each, a total of $7 \times 7 \times 7 = 343$ possible rules result from the combination. To get around this problem, two branches can be combined to create a (PI+D) or (PD+I) controller. Numerous

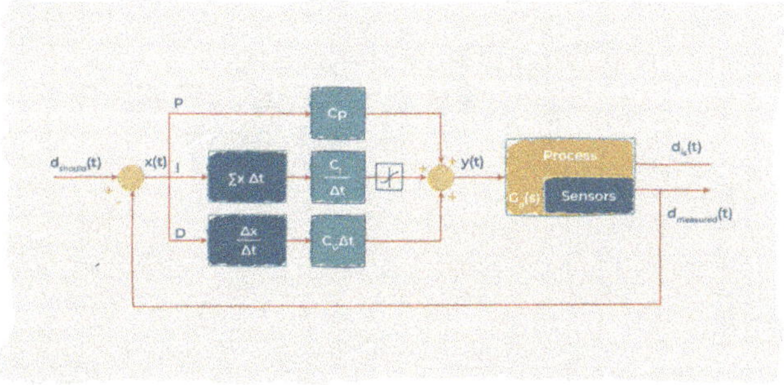

Fig. 8.27 Block diagram of a conventional discrete PID controller

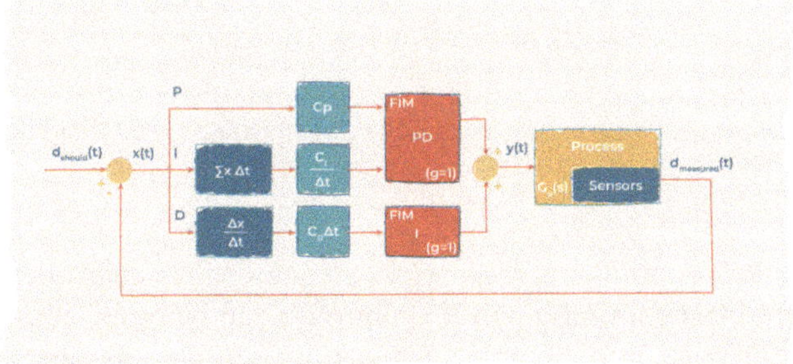

Fig. 8.28 Block diagram of a fuzzy (PD+I) controller

publications indicate that the (PD+I) solution has better properties. A block diagram is shown in Fig. 8.28.

This configuration results in a total of $7 \times 7 + 7 = 56$ possible production rules and thus a significantly lower computational effort to determine the output value. Just as in Fig. 8.7, the signal is differentiated and integrated before being fed into the fuzzy inference machines, which can be done either numerically or directly in the analog signal. The conditions for functional similarity is that the gain factors g of both fuzzy inference machines are approximately equal to 1.

When developing such a fuzzy (PD+I) controller, you can start with the PD part, for example. However, just like a conventional PD controller, this produces an offset error that cannot be corrected. Therefore, the integral part is connected in parallel. The additional summation of the control error leads to the desired output setpoint $d_{\text{measured}} = d_{\text{set}}$. A transfer element with a nonlinear characteristic curve connected in series to the integral term can significantly improve the robustness of the controller.[5] Fig. 8.29 shows examples of the results of position control of a single-jointed robot using a conventional and a fuzzy PID controller.

---

[5] An experimentally motivated, detailed description of our own experiences in designing a fuzzy PID controller can be found, for example, in Brubaker (1994).

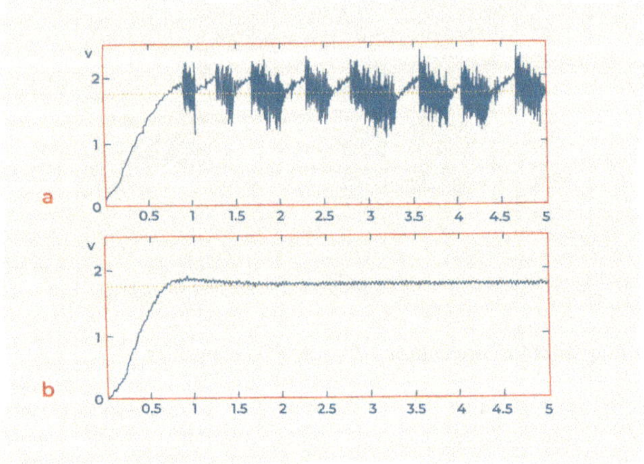

Fig. 8.29 Step response of a conventional controller (**a**) and a fuzzy PID controller (**b**) (according to Malki et al., 1994)

If the fuzzy PID controller is regarded as an engine map controller , the possible improvements over a conventional PID controller can be easily illustrated. Due to the significant increase in freely selectable parameters (i.e., membership functions, rules), the actual engine map can in most cases be better adapted to the engine map that is optimal for the process, which can be very complex depending on the nonlinearity of the process. The advantage over other engine map controllers is the direct interpretability of the parameters and thus an often greatly simplified controller design .

In general, the strengths of fuzzy controllers lie particularly in the control of nonlinear processes, even if these contain delay or dead time elements, in their often significantly greater robustness to parameter changes within the process, and in the very simple implementability of panic rules.

References

Akahoshi, T. (1991). Fuzzy reasoning and its application to control systems. *Fuzzy Sets and Systems, 43*(3), 315–326.

Bandemer, H. (1990). *Fuzzy data analysis*. Springer.

Bothe, H.-H. (1995). *Bewertung mit unscharfen Mengen*. Technische Hochschule Ilmenau.

Brinkmann, R., & Moraga, C. (1992). Fuzzy implications and generalized modus ponens. *Fuzzy Sets and Systems, 48*(1), 41–56.

Brubaker, T. A. (1994). Analysis and design of fuzzy control systems. *IEEE Control Systems Magazine, 14*(3), 45–53.

Buckley, J. J. (1993). Fuzzy probabilities. *Fuzzy Sets and Systems, 55*(2), 133–142. https://doi.org/10.1016/0165-0114(93)90206-R

Cannon, W. B. (1966). *The wisdom of the body*. W. W. Norton.

Driankov, D., Hellendoorn, H., & Reinfrank, M. (1993). *An introduction to fuzzy control*. Springer.

Einstein, A. (1921). *Geometry and experience* [Talk. Prussian Academy of Sciences, Berlin, Germany].

Furuta, K., Yamakawa, T., & Miki, T. (1984). Fuzzy control of industrial processes. *International Journal of Control, 39*(4), 787–801.

Gariglio, S. (1990). *Fuzzy decision models and evaluation methods* [Reference cited in context of fuzzy decision making; exact publication details to be verified].

Hartmann, K. (1992). *Fuzzy set theory-applications in the engineering sciences*. Springer.

Hayashi, I. (1991). Application of fuzzy neural networks to pattern classification. *IEEE Transactions on Systems, Man, and Cybernetics, 21*(5), 1273–1281.

He, X., Wang, L., & Sun, Y. (1993). Fuzzy adaptive control for nonlinear systems. *Fuzzy Sets and Systems, 58*(2), 205–215.

Holmblad, L. P., & Østergaard, J. J. (1982). Control of a cement kiln by fuzzy logic. *Information and Control, 52*(2), 146–170. https://doi.org/10.1016/S0019-9958(82)80012-9.

Kickert, W. J. M. (1975). *Fuzzy theories on decision making*. Martinus Nijhoff.

Landgraf, R. (1970). *Regelungstechnik*. Springer.

Lee, C.-T. (1992). Fuzzy logic in control systems: Fuzzy logic controller—part i. *IEEE Transactions on Systems, Man, and Cybernetics, 20*(2), 404–418.

Malki, H. A., Li, H., & Chen, G. (1994). New design and stability analysis of fuzzy control systems. *IEEE Transactions on Fuzzy Systems, 2*(4), 245–254.

Mamdani, E. H. (1975). An experiment in linguistic synthesis with a fuzzy logic controller. *International Journal of Man-Machine Studies, 7*(1), 1–13. https://doi.org/10.1016/S0020-7373(75)80002-2.

Norita, H. (1992). Design of fuzzy controllers based on linguistic models. *Fuzzy Sets and Systems, 49*(2), 193–204.

Pedrycz, W. (1989). *Fuzzy control and fuzzy systems*. John Wiley & Sons.

Preuß, R. (1992). *Fuzzy-control*. VEB Verlag Technik.

Sturgeon, W. C., & Loscutoff, J. (1972). Stability and control of fuzzy systems. *IEEE Transactions on Systems, Man, and Cybernetics, 2*(2), 194–200.

Yasunobu, S., & Miyamoto, S. (1985). Automatic generation of fuzzy control rules by self-organizing learning. *Fuzzy Sets and Systems, 15*(3), 237–252.

Zimmermann, H.-J. (1991). *Fuzzy set theory — and its applications* (2nd ed.). Kluwer Academic Publishers.

Chapter 9
Fuzzy Pattern Recognition

Abstract This chapter introduces fuzzy pattern recognition, a branch of artificial intelligence that uses fuzzy logic to identify and classify data that is imprecise and noisy or lacks sharp boundaries. It allows for partial membership, meaning an object can belong to multiple classes simultaneously to varying degrees.

Pattern recognition methods are used in all areas of science and technology (e.g., see Bothe, 1995). To illustrate the broad scope of this field, examples include automatic ECG analysis and pulse monitoring, computer tomography, weather forecasting, industrial quality control, process identification, and computer vision in robotics. Thereby, speech and image recognition are particularly important areas.

The objects to be recognized are first represented in an idealization process using characterizing features and then assigned to specific representative patterns (clusters). These clusters are either determined by an expert using specific prior knowledge or formed automatically on the basis of a sample set of objects.

A general procedure for pattern recognition is shown in the block diagram in Fig. 9.1.

The objects of the physical world are abstracted and transferred to the measurement data space by means of measurements. This results in measurement data vectors $\mathbf{x}$, from which suitable characteristic features are to be extracted. This is done by abstracting the objects (i.e., by modeling). After feature extraction, the objects are represented by feature vectors $\mathbf{m}$ in a feature space. If fixed feature classes (or clusters) Q are specified, the objects can be grouped together by "classification." The feature classes form the classification space. If they are generated in an automatic process, this is referred to as automatic "cluster formation" or "clustering." In this case, a representative sample set can be used in a learning phase. If cluster transformations take place during classification, this is referred to as "learning classification."

By transforming the objects of the physical world into the classification space, the (usually vast) diversity of objects in the physical world is reduced to a limited number of predefined patterns.

H.-H. Bothe, E. Portmann, *Computing with Words*, Fuzzy Management Methods,
https://doi.org/10.1007/978-3-032-24117-7_9

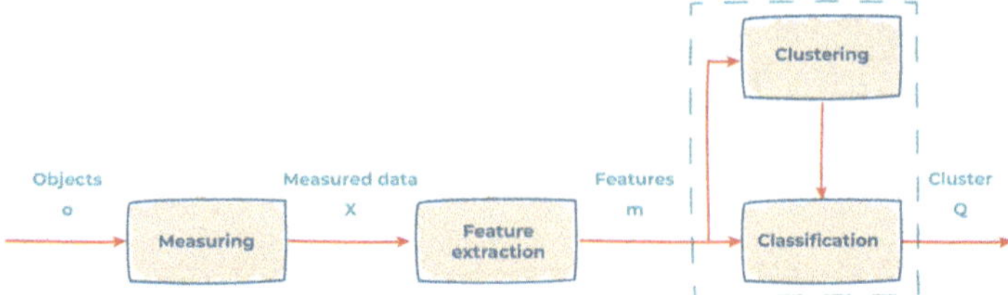

Fig. 9.1 General procedure for pattern recognition

9.1 Object, Feature, and Cluster

In this chapter, the terms "object," "feature," and "cluster" used above will be defined and explained in more detail.

Definition 9.1 (Object) An object is an abstract model of a real system that is defined by a set of measurable model parameters.

Definition 9.2 (Feature) The model parameters according to Definition 9.1 are also called features. These represent characteristic properties of objects and thus have a distinguishing effect. They classify the objects by forming areas in a feature space that is limited by them. A distinction is made between primary features and secondary features derived from them.

Definition 9.3 (Cluster (Class)) The grouping of objects according to a defined similarity criterion (metric) leads to (object) clusters. If the criteria are determined by content, they are referred to as "semantic clusters"; if they are determined by formal criteria, they are referred to as "natural clusters." The process of grouping is called "classification." Clusters can in turn be regarded as objects in a further process.

Example 9.1 In a traffic count, passing vintage motorcycles (our objects) are to be classified. To do this, primary features such as "manufacturer" and "type" are defined in advance, which serve to distinguish between objects. The individual (passing) objects are then assigned to a class such as "NSU150," "DKW 200," or "BMW 500." Obviously, in this case, a clear classification into natural classes is possible. However, based on the data obtained, semantic classes can also be formed, such as "mid-range" and "luxury class." If the two secondary features $m_1 =$ "cubic capacity" and $m_2 =$ "engine power" are used for this purpose, the distinguishing criterion could (naively) be defined by the diagram in Fig. 9.2:

Another way to form classes is to classify in the one-dimensional feature space of the secondary feature $m_3 = m_1 m_2$.

The use of sharp features is problematic in many cases. This example shows that it can be useful to use fuzzy features (e.g., in our case *elegance* and *sportiness*) for classification. Objects and clusters can be represented with sharply defined boundaries in the following ways, among others:

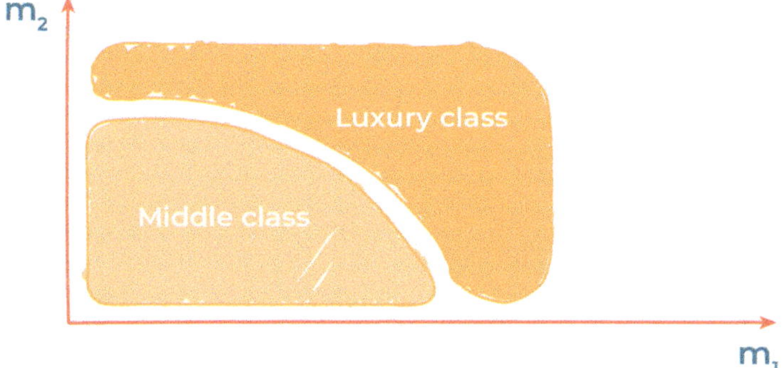

Fig. 9.2 Definition of natural classes for vintage motorcycles

9.1.1   Set Representation

The objects are represented directly by the corresponding feature vectors, the clusters by enumerating the objects; if the license plate is specified as the sole feature, the cluster Q_{Berlin} of vintage motorcycles registered in Berlin could be represented, for example, by:

$$Q_{\text{Berlin}} = \{\text{B–DE8360},\ \text{B–AY1956},\ \text{B–Y2509}, \ldots\}.$$

9.1.2   Analytical Representation

The objects are identified by points in the feature space and grouped by delimitation into clusters. This can be done either by introducing feature-bound intervals (Fig. 9.3a) or by analytically specified separation functions (Fig. 9.3b). This results in point clouds in the feature space.

9.1.3   Logical Representation

The membership of the objects x_i, with the features $m_{i,j}$ to the cluster Q_j, is determined by a logical operator of the $m_{i,j}$; for N objects of Q_j, the rule applies:

$$\text{IF } m_{1,j} \wedge m_{2,j} \wedge \ldots \wedge m_{N,j} \text{ THEN } Q_j. \tag{9.1}$$

In most cases, it makes sense to specify tolerance intervals for the individual features. If these tolerance intervals are evaluated with regard to a possible occurrence, the features can be represented by fuzzy sets and the rules for cluster membership

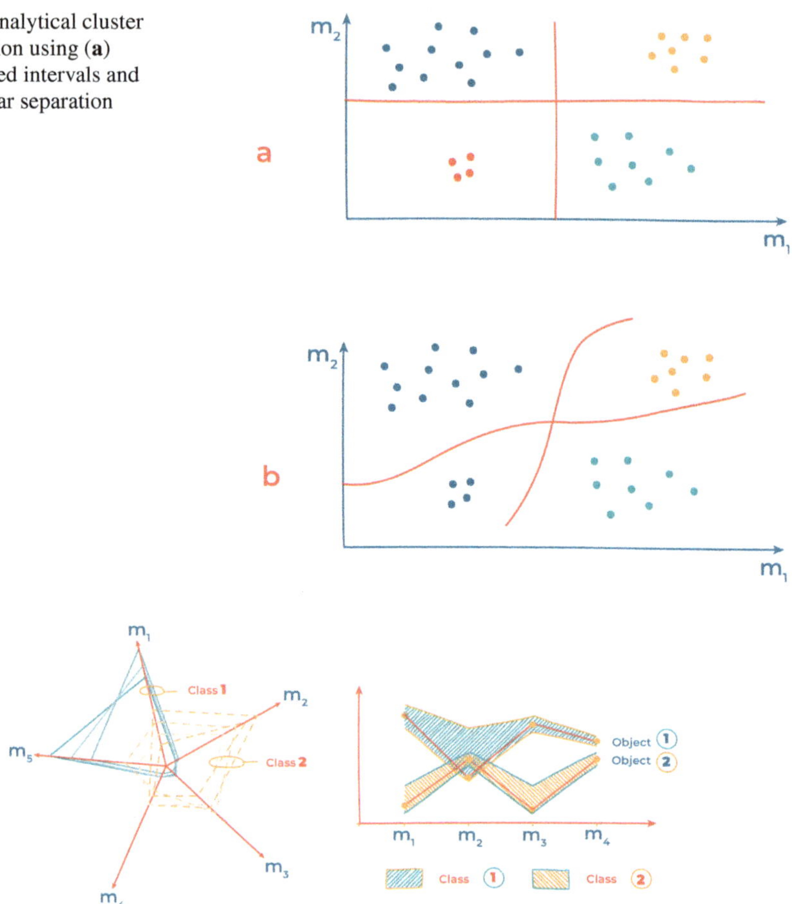

Fig. 9.3 Analytical cluster representation using (**a**) feature-based intervals and (**b**) nonlinear separation functions

Fig. 9.4 Cluster representation: left, polar representation; right, profile representation

by fuzzy relations. The classifier is then constructed in the same way as a fuzzy controller according to Sect. 8.2, and the actual objects are assigned to the clusters Q_j by solving a fuzzy relational equation system.

9.1.4   Graphical Representation

To illustrate point clouds in a feature space with more than three dimensions, a profile representation or a polar representation can be used (Fig. 9.4).

Further representation options can be found, for example, in Bocklisch (1988).

9.2 Fuzzy Classifiers

In the field of pattern recognition , the following approaches to the use of fuzzy logic methods can be found in literature:

1. Allowance of fuzzy cluster boundaries and thus a fuzzy decomposition of the basic domain X; each object belongs to the individual clusters with varying intensity.
2. Allowance of fuzzy feature vectors to characterize the objects; each object belongs to the individual clusters with varying intensity. The clusters can be specified as sharp or fuzzy.
3. Use of fuzzy methods in cluster formation; this can result in fuzzy or sharply defined clusters.

This raises two questions: How is the feature space divided into clusters, and what methods can be used to assign an actual object to a given cluster? In this chapter, we will first assume that the clusters are already given and postpone the first question. To discuss the second question, we will define the term "classifier."[1]

Definition 9.4 (Classifier) A classifier is an algorithm that assigns an unknown object x_i (i.e., characterized by the feature vector m_j) to one or, in the case of fuzzy classifiers, several clusters Q_k. It thus divides the basic set of objects x_i into n subsets.

Definition 9.4 initially refers to the physical world. However, the division of objects by a classifier takes place in automatic (i.e., computer-aided) classification in the feature space. If the individual clusters are represented by membership functions, a classifier maps the actual feature vectors to membership values for the clusters. If fuzzy boundaries are allowed, this results in a fuzzy classifier.

Definition 9.5 (Fuzzy Classifier) A fuzzy classifier is an algorithm that decomposes the feature space X^M into n fuzzy subsets $Q_1, \ldots, Q_n$, where for all $m_i \in X^M$ applies:

$$\sum_{j=1,\ldots,n} \mu_{Q_j}(m_i) = 1.$$

The Q_j are also called fuzzy clusters. Each object with $m_i \in X^M$ is divided precisely among all fuzzy clusters according to its membership values.

The memberships in Definition 9.5 can also be written in a matrix form using the abbreviations $u_{ij} = \mu_{Q_j}(m_i) \ \forall i = 1, \ldots, N, \ j = 1, \ldots, n$. This results in a matrix $\mathbf{U} = (u_{ij})$, for which the following applies:

---

[1] Classifiers can not only be used for classification, but also for automatic clustering .

$$\sum_{j=1,\ldots,n} u_{ij} = 1. \tag{9.2}$$

In addition, the requirement is made that:

$$0 < \sum_{i=1,\ldots,N} u_{ij} < N. \tag{9.3}$$

The summation of the membership values u_{ij} of an object X_i over all n clusters Q_j must equal to 1 according to Definition 9.5 and (9.2). Since the membership values u_{ij} are always positive, the summation of u_{ij} over all objects x_i of a cluster Q_j must lie between 0 and the total number N of objects according to (9.3).

A different representation results when the membership values $\mu_{Q_j}(m_i)$ are arranged into an n-dimensional vector:

Definition 9.6 (Vector of Sympathy) The vector of sympathy $\mu(m_i)$ of an object x_i is defined by the membership of the feature vector m_i to the n clusters Q_j:

$$\mu(m_i) = \left(\mu_{Q_1}(m_i), \ldots, \mu_{Q_n}(m_i) \right).$$

A fuzzy classifier assigns a vector of sympathy $\mu(m_i)$ to each object x_i.

An immediately obvious method for constructing a fuzzy classifier is to fuzzify the feature values (as in Example 1.6); the clusters correspond to the terms of a linguistic variable, and the course of their membership functions determines the vector of sympathy.

An extended method determines the classification rules using IF…THEN…rules and leads to a fuzzy relational equation system. The mode of operation of a fuzzy classifier defined in this way can be directly compared to that of a fuzzy controller (according to Sects. 1.2 or 8.2), whereby no defuzzification is performed initially. The values of the feature vector m_i are at the input of the classifier, and the output values are formed from the sympathy values $\mu_{Q_j}(m_i)$ for the individual clusters Q_j. The value range of the output variables is thus limited to the unit interval $[0, 1]$. With sharp feature values, this results in a situation similar to that in Example 1.6 and, with fuzzy feature values, similar to that in Example 8.4. The fuzziness of the feature values can arise from weighted tolerance specifications (as already shown in Example 8.1). The possible solutions for the resulting fuzzy relational equation system can be taken from Sects. 7.3 and 8.2.

The largest single value $\mu_{\max}(m_i) = \max_j u_{ij}$ of the vector of sympathy $\mu(m_i)$ is called the "main sympathy value" of the object x_i; by specifying a threshold value for $\mu_{\max}(m_i)$, the "identifiability" of the object can be determined by a given classifier. The second-largest sympathy value (largest "secondary sympathy value") can be used to make a risk statement for identification. For this purpose, the term "signal-to-noise ratio" is defined.

Definition 9.7 (Signal-to-Noise Ratio) The signal-to-noise ratio $\Delta\mu(m_i)$ of an object x_i with respect to the n clusters Q_j, $j = 1, \ldots, n$ is defined by:

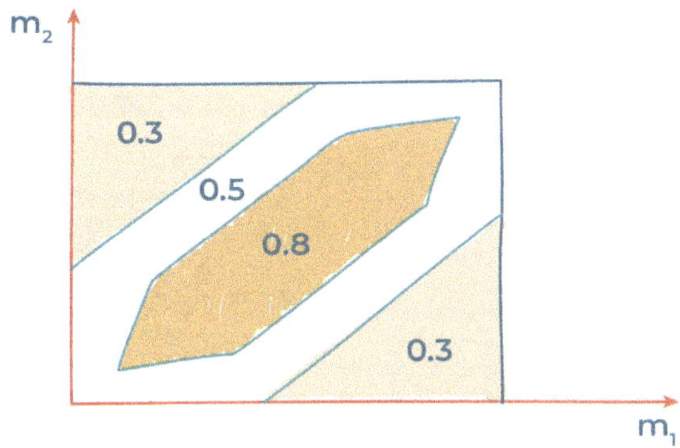

Fig. 9.5 Defining a required signal-to-noise ratio by range

$$\Delta\mu(m_i) = \frac{\mu_{\max}(m_i) - \max_{j,\ j \neq j^*} u_{ij}}{\mu_{\max}(m_i)}.$$

With $\max_{j,\ j \neq j^*}(u_{ij}) < \mu_{\max}(m_i)$, it is obvious that $\Delta\mu(m_i) \in [0, 1]$, whereby optimal classification is characterized by $\Delta\mu(m_i) = 1$. A risk statement is now made by defining ranges in which the signal-to-noise ratio must not fall below certain minimum values. Figure 9.5 shows a corresponding range definition, for example, for a two-dimensional feature space with the features m_1 and m_2.

In principle, the use of several alternative classifiers is conceivable for a specific problem. These can draw on different features or be based on different similarity measures. When selecting a specific classifier, the question arises as to how to evaluate its functionality or quality criteria. This should assign each object to the "correct" cluster based on its main sympathy value, whereby the secondary sympathy values differ significantly from the main sympathy value.

It often seems more advantageous to determine an optimal classifier from a list of possible classifiers rather than specifying a concrete one. To do this, evaluation or quality criteria must be known that can initially be applied to a representative selection of objects. The following criteria are suggested by Bocklisch (1988), for example; the specific choice must be made on a problem-specific basis and is therefore left to the intuition and skill of the developer. Once a quality criterion has been defined, the process of classifier selection can be automated.

9.2.1 Quality Criteria for Selecting Fuzzy Classifiers

Let there be a sample of objects $\underline{x}_i$ with $i = 1, \ldots, I$ that is representative of the problem. Then the quality criteria g_1 to g_6 can be used to select an optimal classifier:

1. The main sympathy values $\mu_{\max} * (\underline{m}_i)$ of the individual sample objects $\underline{x}_i$ are used that would classify them in the "correct" cluster; this defines the relative quality measures g_1 and g_2:

$$g_1 = (1/I) \sum_{i=1,\ldots,I} \mu_{\max} * (\underline{m}_i) \quad \text{and} \tag{9.4}$$

$$g_2 = \frac{\sum_{i=1,\ldots,I} \mu_{\max} * (\underline{m}_i)}{\sum_{i=1,\ldots,I} \mu_{\max}(\underline{m}_i)}. \tag{9.5}$$

The values of g_1 and g_2 should be close to 1.

2. The mean secondary sympathy value or the maximum secondary sympathy value relative to the main sympathy value should be small. To meet this requirement, the object-specific terms

$$g_{3,i} = \frac{\left[\sum_{j=1,\ldots,n} u_{ij}\right] - \mu_{\max}(\underline{m}_i)}{(n-1)\,\mu_{\max}(\underline{m}_i)} \tag{9.6}$$

and

$$g_{4,i} = \max_{j=1,\ldots,k-1,k+1,\ldots,n} \left[\frac{u_{i,j}}{\mu_{\max}(\underline{m}_i)}\right] \tag{9.7}$$

are formed and averaged over the sample objects. In case (9.7), the value k denotes the cluster to which the main sympathy value belongs; the maximum is formed only over the secondary sympathy clusters.

When using the arithmetic mean across all sample objects $\underline{x}_i$, the quality measures are

$$g_3 = 1/I \sum_{i=1,\ldots,I} g_{3,i} \tag{9.8}$$

and

$$g_4 = 1/I \sum_{i=1,\ldots,I} g_{4,i}. \tag{9.9}$$

The use of a mean value of the main sympathy values of all sample objects $\underline{x}_i$ weighted by the factor $g_{3,i}$ leads to:

$$g_5 = \frac{\sum_{i=1,\ldots,I}\left[\left(\sum_{j=1,\ldots,n} u_{ij}\right) - \mu_{\max}(\underline{m}_i)\right]}{\sum_{i=1,\ldots,I} \mu_{\max}(\underline{m}_i)}. \tag{9.10}$$

In this case, we also refer to the "degree of coverage." Using the weighting factor $g_{4,i}$ results in the quality measure:

$$g_6 = \frac{\sum_{i=1,\ldots,I} \max_{j=1,\ldots,k-1,k+1,\ldots,n} \left(u_{i,j}\right)}{\sum_{i=1,\ldots,I} \mu_{\max}(\underline{m}_i)}. \tag{9.11}$$

The measures g_4 and g_6 place "stricter" demands on the classifier than the measures g_3 and g_5, since, especially with a larger number of features, the mean value of the secondary sympathy values can be considerably smaller than the largest secondary sympathy value.

9.3 Automatic Clustering

The clusters are either specified by an expert (e.g., "logical representation" in Sect. 9.1) or formed automatically. In the second case, the following problem arises:

Let $X^M = \{\underline{x}_1, \ldots, \underline{x}_N\}$ be a set of N distinguishable objects $\underline{x}_j$ each with M features. X^M is to be divided into subsets in such a way that objects belonging together are as similar as possible to each other, while objects from different subsets are as dissimilar as possible.

For automatic clustering, a clear definition of the term "similarity" and thus a measure of similarity must therefore be established. This definition is based on the features, as these have an object-ordering structure. We want to define "similarity" using the term "distance" and will first develop numerical distance measures in the following.[2] A smaller distance then means greater similarity.

Definition 9.8 (Distance) A distance is a sharp distance function $d : \mathbb{R}^M \times \mathbb{R}^M \to \mathbb{R}^+_0$ in the M-dimensional space of feature vectors $\underline{m}$, $\underline{m}'$, and $\underline{m}''$ that satisfies the following conditions:

1. $d(\underline{m}, \underline{m}) = 0$ $\qquad\qquad\qquad\qquad\qquad \forall\, \underline{m} \in \mathbb{R}^M,$
2. $d(\underline{m}, \underline{m}') = d(\underline{m}', \underline{m}) \geq 0$ $\qquad\qquad\quad \forall\, \underline{m}, \underline{m}' \in \mathbb{R}^M,$
3. $d(\underline{m}, \underline{m}'') \leq d(\underline{m}, \underline{m}') + d(\underline{m}', \underline{m}'')$ $\qquad \forall\, \underline{m}, \underline{m}', \underline{m}'' \in \mathbb{R}^M.$

Note If fuzzy features are specified, a corresponding fuzzy distance function **d** can be calculated using the extension principle described in Sect. 3.2.

Here are some examples of distance measures for crisp features. The class of L_p distances is most commonly used.

Definition 9.9 (L_p Distance) Let M be the dimension of the feature space. Then the class of L_p distances $d_{ij}{}^L$ between two objects $\underline{x}_i, \underline{x}_j \in X$ with the feature vectors $\underline{m}_i, \underline{m}_j \in \mathbb{R}^M$ is given by:

$$d_{ij}{}^P = d^P(\underline{m}_i, \underline{m}_j) = \left[\sum_{k=1,\ldots,M} \left|m_{ik} - m_{jk}\right|^p\right]^{1/p} \qquad \text{with } p > 0.$$

---

[2] Incidentally, these can be used for both clustering and classifying individual objects. A concrete example of a classifier based on distance measures is presented in Sect. 9.4.

The double indexing within the modulus of absolute value is intended to distinguish between the feature vectors and their elements (k). All M features of $\underline{m}_i$ and $\underline{m}_j$ are taken into account in the summation.

In particular, the following special cases can be specified for L_p distances:

Hamming distance:

$$p = 1: \quad d_{ij}{}^H = \sum_{k=1,\dots,M} |m_{ik} - m_{jk}|, \tag{9.12}$$

Euclidean distance:

$$p = 2: \quad d_{ij}{}^E = \left[\sum_{k=1,\dots,M} (m_{ik} - m_{jk})^2\right]^{1/2}, \tag{9.13}$$

Greatest axis-parallel distance:

$$p \to \infty: \quad d_{ij}{}^C = \max_{k=1,\dots,M} |m_{ik} - m_{jk}|. \tag{9.14}$$

If the individual axes of the feature space are to be weighted differently (e.g., if certain features appear to be more important than others), a weighting factor w_k can be introduced within the summation for the individual axis-parallel distances. The L_p distances are thus extended.

Definition 9.10 (Minkowski Distance) Let M be the dimension of the feature space. Then the class of Minkowski distances $d_{ij}{}^P$ of two objects $\underline{x}_i, \underline{x}_j \in X$ with the feature vectors $\underline{m}_i, \underline{m}_j \in \mathbb{R}^M$ is given by:

$$d_{ij}{}'^P = d'^P(\underline{m}_i, \underline{m}_j) = \left[\sum_{k=1,\dots,M} w_k |m_{ik} - m_{jk}|^P\right]^{1/p} \quad \text{with } p > 0.$$

As in Definition 9.9, the double indexing within the modulus of absolute value is intended to distinguish between the feature vectors and their elements (k), where w_k represents the weighting factors.

In particular, the following special cases can be specified for Minkowski distances:

City block distance:

$$p = 1: \quad d_{ij}{}'^H = \sum_{k=1,\dots,M} w_k |m_{ik} - m_{jk}|, \tag{9.15}$$

Weighted Euclidean distance:

$$p = 2: \quad d_{ij}{}'^E = \left[\sum_{k=1,\dots,M} w_k |m_{ik} - m_{jk}|^2\right]^{1/2}, \tag{9.16}$$

Largest weighted axis-parallel distance:

$$p \to \infty : \quad d_{ij}^{\,\prime C} = \max_{k=1,\dots,M} w_k \left| m_{ik} - m_{jk} \right| . \tag{9.17}$$

Definition 9.11 (Kendall's Rank Distance) The objects $\underline{x}_i$ are first ordered by features using rank numbers. Overall, the feature vectors $\underline{m}_i = (m_1, \dots, m_M)_i$ are mapped to corresponding rankings $\underline{r}_i$ of the x_i, i.e., to simple number sequences. For each object x_i, the following applies:

$$\underline{m}_i = (m_1, \dots, m_M)_i \to \underline{r}_i = (r_1, \dots, r_M)_i,$$

where $i = 1, 2, \dots, n$ and $r_i = 1, 2, \dots, n$. The rank distance $d_{ij}^{\,r}$ between two objects x_i and x_j is then defined by:

$$d_{ij}^{\,r} = 1/(n^2 - 1) \sum_{k=1,\dots,M} \left(r_{ki} - r_{kj} \right)^2 .$$

where r_{ki} and r_{kj} are the kth ranks of the objects $\underline{x}_i$ and $\underline{x}_j$.

The feature values themselves no longer play a role in determining the rank distance; only the resulting ranking of the objects is relevant. The rank distance is therefore scale-invariant (i.e., it does not depend on any nonlinearity or offset of the measuring range scale). The ranking can be determined, for example, by the size of the respective main sympathy values or a combination with the secondary sympathy values.

9.4   Determination of Fuzzy Clusters

Automatic clustering is usually based on a sample set of representative objects. However, if a closed dataset is available, it can also be performed on the total set of objects to be classified, so that no sample selection is necessary. In the following, we will generally assume a number N of objects, which may correspond to a sample set in a dynamic classification task or otherwise to the total set of objects.

The most common method of cluster formation is probably hierarchical clustering, which is described, for example, in Bock (1974) and Späth (1975). Clustering is performed here by calculating the local distance between the feature vectors of the objects. Given a threshold value d_s, those objects that are close enough to each other are grouped into clusters. Variations in d_s result in different cluster configurations, which can be represented using dendrograms, as shown in Fig. 9.6. When d_s is specified according to the dashed line, the clusters $\{1, 2, 3, 4, 5, 6\}$ and $\{7, 8, 9, 10\}$ are formed.

An elegant way to automatically form cluster centers using fuzzy sets is described by Gitman and Levine (1970). It consists of interpreting the relative object frequency in a δ-environment (or distance) d_{ij} of the points $\underline{m}_i$ of the feature space as the

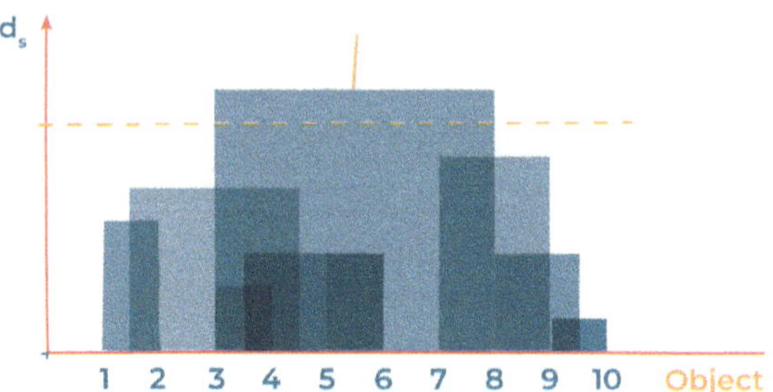

Fig. 9.6 Dendrogram illustrating hierarchical clustering

membership value of a fuzzy set $\mathbf{A}$ on $\mathbb{R}^M$; the maxima of $\mu_A(\underline{m}_i)$ then determine the centers $\underline{v}_j$ of the clusters sought as follows:

$$\underline{v}_j = 1/N \, \text{card}\,[\underline{m}_i \in \mathbb{R}^M \mid d_{ij} < \delta]; \tag{9.18}$$

where $\text{card}\,[\underline{m}_i \in \mathbb{R}^M \mid d_{ij} < \delta]$ is a function that returns the number of feature vectors that satisfy the condition as its function value. The cluster boundaries can be determined on the basis of a specific distance measure in a first approximation by means of a sharp, symmetrical division of the area.

If the clusters are specified in an initial approximation (e.g., using the method described above or manually), they are likely to be filled relatively inhomogeneously by the objects. The position of the clusters can then be optimized in an iterative process according to the objectives described above. To do this, the new cluster centers can first be estimated by calculating a weighted centroid $\underline{V} = (\underline{v}_1, \ldots, \underline{v}_n)^T \in \mathbb{R}^M$, and then the objects can be reassigned. For $\underline{V}$, the following applies:

$$\underline{v}_j = \frac{\sum_{i=1,\ldots,I} [u_{i,j}]^q \, m_{ij}}{\sum_{i=1,\ldots,I} u_{i,j}}, \quad j = 1, \ldots, n, \quad q \in [1, \infty], \tag{9.19}$$

where $\underline{U} = (u_{i,j})$ is calculated, for example, from the distances of the objects $\underline{x}_i$ to the old cluster centers. This calculation is performed using a function f that maps the value range of d_{ij} reciprocally to the unit interval $[0, 1]$. A simple analytical function can be specified with $f(d_{ij}) = [1 + d_{ij}]^{-1}$. $\underline{U}$ can also initially represent a sharp cluster assignment of $\underline{x}_i$ with $u_{i,j} \in \{0, 1\}$, as long as it satisfies the conditions (9.3).

The initial solution needs to be further optimized. A suggestion by Bezdek (1981) leads with $q \in [1, \infty]$ to the minimization of the functional

$$G(\underline{U}, \underline{V}) = \sum_{i=1,\ldots,N} \sum_{j=1,\ldots,n} (u_{ij})^q \, d^2(\underline{m}_i, \underline{v}_j) \tag{9.20}$$

with respect to the matrices $\underline{U}$ and $\underline{V}$; here, $\underline{v}_j$ represents the cluster centers, and u_{ij} represents the membership values of $\underline{x}_i$ to the clusters Q_j. Minimization with respect to $\underline{V}$ tends to optimize the cluster centers, while minimization with respect to

$\underline{U}$ tends to optimize the object assignment. The parameter q can be freely selected and determines the cluster fuzziness; as q increases, the clusters tend to become fuzzier. For this reason, q is also called the "contrast parameter."

Example 9.2 (ISODATA-FCM Method) One possible minimization method for the functional $G(\underline{U}, \underline{V})$ according to (9.20) is the ISODATA-FCM iteration method according to Bezdek (1981); it consists of four individual steps, converges to a local minimum, and is described below:

Step 1: Select

-    The desired number of clusters n with $2 \leq n \leq N$
-    A suitable distance measure such as the Euclidean distance
-    A suitable contrast parameter q
-    The initial assignment $\underline{U}^{(0)} = (u_{ij})^{(0)}$ of the membership matrix of the objects $\underline{x}_i$ to the clusters Q_j with:

$$\sum_{j=1,\ldots,n} u_{ij} = 1 \quad \text{and} \quad 0 < \sum_{i=1,\ldots,N} u_{ij} < N.$$

Step 2 Let s be the number of current iteration steps. Calculate the n cluster centers $\underline{v}_j{}^{(s)}$ according to:

$$\underline{v}_j{}^{(s)} = \frac{\sum_{i=1,\ldots,N}\left(u_{ij}{}^{(s)}\right)^{q*} \underline{m}_i}{\sum_{i=1,\ldots,N}\left(u_{ij}{}^{(s)}\right)^{q}}.$$

Step 3 Calculate $\underline{U}^{(s+1)} = (u_{ij}{}^{(s+1)})$. To do this, let

$$I_i = \{\, j \in \{1, \ldots, n\} \mid d(\underline{m}_i, \underline{v}_j) = 0 \,\} \quad \text{and}$$
$$I_i^c = \{1, \ldots, n\} \setminus I_i.$$

I$_i$= 0 :

$$\Rightarrow \quad u_{ij}{}^{(s+1)} = \frac{1}{\sum_{k=1,\ldots,n} d(\underline{m}_i, \underline{v}_j)^{(s)} / d(\underline{m}_i, \underline{v}_k)^{(s)}} \,,$$

I$_i$ $\neq$ 0 :

$$\Rightarrow \quad u_{ij}{}^{(s+1)} = 0 \ \text{ for all } \ j \in I_i^c$$

and subsequent renormalization according to:

$$\sum_{j \in I_i} \left(u_{ij}{}^{(s+1)}\right) = 1.$$

Step 4 Terminate if

$$\left\| \underline{U}^{(s+1)} - \underline{U}^{(s)} \right\| \le \varepsilon$$

with $\varepsilon > 0$ and $\| \cdot \|$ a matrix norm (e.g., the Euclidean norm) matching d_{ij}.

In deviation from the termination criterion according to Step 4, in many cases the iteration can also be carried out until the membership values of the objects to the individual clusters remain essentially constant (i.e., with sufficiently large signal-to-noise ratios, the cluster memberships of the main sympathy values of the objects remain constant).

References

Bezdek, J. C. (1981). *Pattern recognition with fuzzy objective function algorithms.* Springer.

Bock, H. H. (1974). *Automatische Klassifikation.* Vandenhoeck & Ruprecht.

Bocklisch, F. (1988). *Prozessanalyse mit unscharfen Verfahren* VEB Verlag Technik. Berlin.

Bothe, H.-H. (1995). *Bewertung mit unscharfen Mengen.* Technische Hochschule Ilmenau.

Gitman, I., & Levine, M. D. (1970, July). An algorithm for detecting unimodal fuzzy sets and its application as a clustering technique. *IEEE Transactions on Computers, C-19*(7), 583–593. https://doi.org/10.1109/T-C.1970.222992

Späth, H. (1975). *Clusteranalyse: Algorithmen zur Objektklassifizierung und Datenreduktion.* Oldenbourg.

Chapter 10
Fuzzy Hardware Implementation

Abstract This chapter shows fuzzy hardware implementations, which are physical circuits designed to process fuzzy logic directly on dedicated hardware, rather than emulating it through software on conventional processors.

When simulating fuzzy logic on conventional digital computers, the linking operators must be replicated using programming instructions. Real-time processing easily leads to very high demands on the computing speed of the hardware. For example, if a fuzzy controller is specified in accordance with Sect. 8.2 with two input variables, one output variable, and a rule base with seven rules, depending on the type of microprocessor used, approximately 20,000 cycle times must be calculated in order to calculate the control value; the cycle time should indicate the average processing time of a simple assignment (Bothe, 1995). Although modern processors can achieve this computing speed easily, especially in parallel operation, their use is not always advisable because many of the implemented functions remain unused. This results in a poor price-performance ratio.

Instead, special hardware structures are advantageous for the economical use of fuzzy logic (especially in consumer electronics like wearable technologies),[1] where hardware costs account for a high proportion of the total costs. For larger projects, these may be special fuzzy processors or controllers that have been developed specifically for solving fuzzy methods and work very quickly on the basis of a corresponding hardware design. This allows fuzzy operations (e.g., intersection, union, and negation) to be performed, multiple rules to be processed in parallel, and fuzzification and defuzzification to be implemented directly in hardware.

In many cases, however, the use of such programmable components is too costly; if the implementations for the linguistic operators, implication operators, etc. are fixed (e.g., in small consumer electronics devices), the operations can also be established with the aid of fuzzy gate logic. Needless to say, the user-specific circuits developed can also be combined into highly integrated components.

---

[1] By wearable technology we mean consumer electronics designed to be worn on the human body, incorporated into gadgets, accessories, or clothes.

© The Author(s), under exclusive license to Springer Nature Switzerland AG 2026   159
H.-H. Bothe, E. Portmann, *Computing with Words*, Fuzzy Management Methods,
https://doi.org/10.1007/978-3-032-24117-7_10

Both digital and analog circuit technologies can be used to implement the structures. Analog circuit technology appears to be very promising for the hardware implementation of fuzzy methods because it is potentially fast and powerful. It promises very high computing power and direct processing of analog values such as sensor signals without the use of analog-digital converters. In addition, high accuracy is possible even when using a small number of transistors. The disadvantages compared to a digital implementation are a higher susceptibility to interference and more difficult programmability, as is known from analog computers.

Another development goal could be hybrid analog-digital circuits in which the fuzzy operations are analog, but the interfaces (e.g., a synchronous system clock, etc.) are digital.

The following chapters discuss the design of a fuzzy controller in analog circuit technology as an example. For this purpose, reference is made to research work by Yamakawa and Miki (1986), Heite et al. (1989), Watanabe et al. (1990), Zhijian and Hong (1991), and Kettner et al. (1992). The individual circuits are intended as very large-scale implementations in integrated components; however, they can also be replicated with discrete components if the transistors and resistors are carefully selected. In particular, transistor and resistor arrays should then be used for the current mirror circuits. Only self-locking MOS field-effect transistors (MOSFETs) are used, with additional NPN bipolar transistors for defuzzification.[2]

10.1   Current Mirror Circuits

In current-controlled circuits, the membership values $\mu(x)$ are represented by currents, making it very easy to form sums and differences. First, a working range $\Delta I = [I_{\min}, I_{\max}]$ is defined. Through normalization according to

$$i = (I - I_{\min})/I_{\max}, \tag{10.1}$$

a value range $0 \leq i \leq 1$ is defined so that the normalized currents i directly correspond to the membership values μ.

The n-MOS current mirror shown in Fig. 10.1 reproduces an impressed input current i_e as an output current i_a, mirrored at the negative operating voltage U_{SS}. The simplified representation of the n-MOSFETs in the circuit diagram is intended to indicate that the semiconductor substrate is electrically connected to the source terminal.

---

[2] Mimicking brains to build more efficient, low-power hard- and software, metal oxide semiconductors (MOS) are used to create brain-inspired structures; MOS field-effect transistors (MOSFET) therein act as functional elements emulating neurons and synapses; and n- as well as p-types are electrical charges, with electrons for n-types (positive, to turn on) and holes for p-types (negative, to turn off).

Fig. 10.1 Schematic diagram
of an n-MOS current mirror

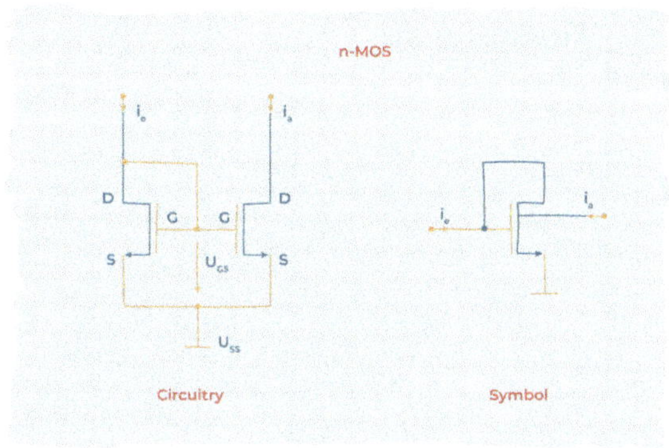

Fig. 10.2 Schematic diagram
of a p-MOS current mirror

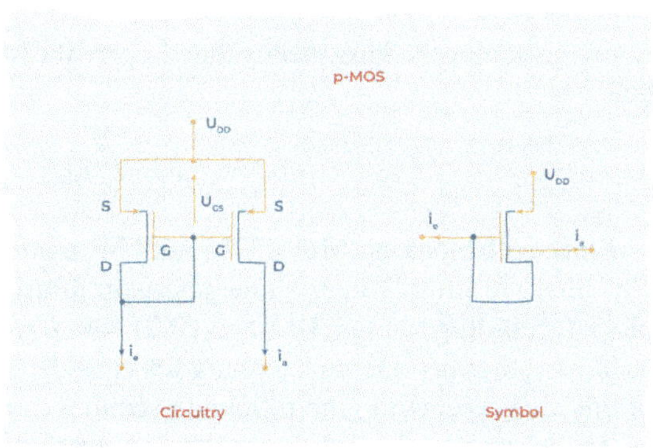

The input current i_e is applied to the left transistor as drain current I_D. The gate-source voltage U_{GS} can be read from the transfer characteristic $I_D = f(U_{GS})$. Since U_{GS} is also applied to the gate-source path of the right transistor, $i_a = i_e$ if both transistors have the same slope S ($I_D = S\,U_{GS}$). When constructing discrete circuits, the two source connections should be connected to the operating voltage U_{SS} in a negative feedback circuit via resistors of equal size. By varying the resistance ratio, a mirror factor $k = i_a/i_e$ can be set. For the n-MOS current mirror with a mirror factor $k = 1$, the short symbol shown in Fig. 10.2 is introduced.

If the n-MOS transistors are replaced by p-MOSFETs and the negative operating voltage U_{SS} is swapped with a positive U_{DD}, a p-MOS current mirror is created as shown in Fig. 10.2. The current directions are reversed; the input current i_e is now mirrored at U_{DD}. The specified symbol for $k = 1$ is introduced.

In a design with discrete components, the two source connections should each be connected to the operating voltage U_{DD} via a resistor to stabilize the mirror factor. It should also be noted that with an n-MOS current mirror, positive input current and output current flow into the circuit, while with a p-MOS current mirror, they flow out of the circuit.

With MOS current mirrors, a set mirror factor remains constant over a wide modulation range and down to $i_e = 0$. Relative deviations of less than 1% can

Fig. 10.3 Schematic diagram
of an n-MOS dual current
mirror

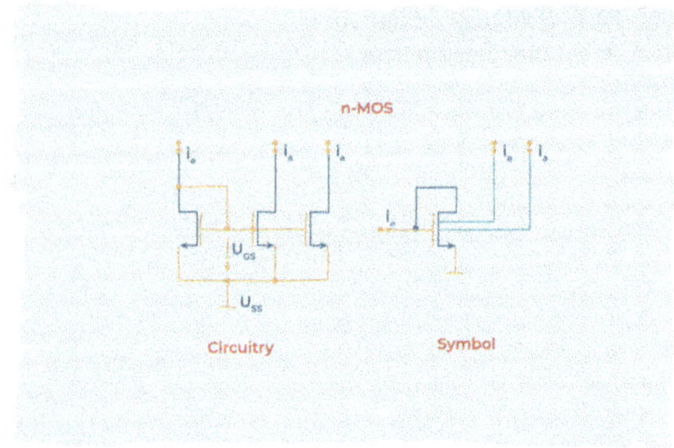

easily be achieved. This means that $I_{\min} = 0$ can be set, and from (10.1) it follows
that:

$$i = I/I_{\max}. \tag{10.2}$$

Since self-blocking MOSFETs only become conductive at gate-source voltages
U_{GS} that are greater than a characteristic threshold value U_T, the current mirrors
block (according to Figs. 10.1 and 10.2) with $U_{GS} < U_T$ for $i_e \leq 0$. This property is
used in numerous operator circuits for sign-dependent decision-making. Examples
of this are presented in the following sections.

These circuits also require multiple current mirrors that mirror the input current
i_e to several equal output currents i_{a1}, i_{a2}, ... An example of a dual current mirror
with the corresponding symbol is shown in Fig. 10.3.

By adding further output transistors with parallel connection of the gate-source
paths, the output current i_a can be reproduced multiple times. Only at high
frequencies do the gate-source capacitances have to be taken into account, so the
propagation delay increases with the number of outputs. If the gate currents are no
longer negligible, the mirror factor also changes.

10.2   Membership Functions

In fuzzy controllers, analog input signals i_{IN} are first fuzzified using membership
functions. This raises the question of how these function curves can be represented.
If input signals and membership values are represented by currents, a current conversion must be performed. This process can be done with the help of current mirrors,
whereby the actual function curve is approximated by straight-line segments (see
Fattaruso & Meyer, 1987). The basic circuits generate linear functions with a
positive slope and negative axis spacing or with a negative slope and positive axis
spacing. They are shown in Fig. 10.4a and b.

Fig. 10.4 Basic circuits for linear function segments with (**a**) a positive slope and negative axis spacing and (**b**) negative slope and positive axis spacing (according to Kettner et al., 1992)

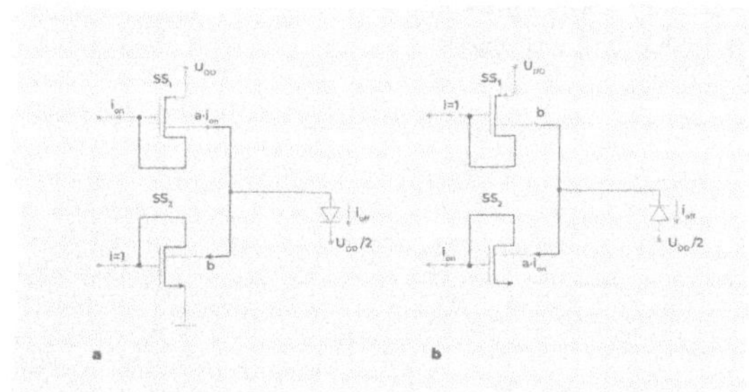

Both circuits use current mirrors with mirror factors of $1 : a$ and $1 : b$ to form straight-line segments of the form:

$$\mu(i_{IN}) = i_{OUT} = \alpha\, i_{IN} + \beta, \qquad \alpha, \beta = \text{const.} \tag{10.3}$$

To generate the straight-line slope a, the input current i_{IN} is mirrored by a factor a; the axis spacing b of the straight line is created by mirroring a reference current $i_{REF} = 1$ by a factor b. As long as $a\, i_{IN} < b\, i_{REF}$ applies, the diode is blocked, and thus $i_{OUT} \approx 0$. For $a\, i_{IN} > b\, i_{REF}$, the output current i_{OUT} is:

$$i_{OUT} = a\, i_{IN} - b\, i_{REF}. \tag{10.4}$$

In a similar way, the following applies to the circuit shown in Fig. 10.4b:

$$i_{OUT} = -a\, i_{IN} + b\, i_{REF}. \tag{10.5}$$

By suitably combining the two basic circuits, piecewise linear functions can now be generated that approximate any membership curves. The larger the number of segments selected, the smaller the approximation error. On the other hand, the overall circuit becomes more complex and occupies a larger chip area.

In order to obtain a continuous function curve, the coefficients α and β of neighboring segments (i) and $(i + 1)$ must satisfy certain conditions. With the corresponding mirror factors a_i, a_{i+1}, b_i and b_{i+1}, the continuity requirement with $\alpha_1 = a_1, \beta_1 = b_1$, and $i_{REF} = 1$ leads to the recursion formulas:

$$a_i = \alpha_i - \alpha_{i-1} \qquad \forall\, i = 2, \ldots, n \quad 1\,. \tag{10.6}$$

$$a_i = \beta_i - \beta_{i-1} \qquad \forall\, i = 2, \ldots, n - 1\,. \tag{10.7}$$

Figure 10.5 shows the implementation of a left-side shoulder function, as used, for example, to mark the term *very dark* of the linguistic variable "brightness" in Example 4.1. To define the function curve, a basic function (according to Fig. 10.4b) and an additional constant current source were used (i.e., the circuit of which can be

Fig. 10.5 Circuit principle
for generating a shoulder
function

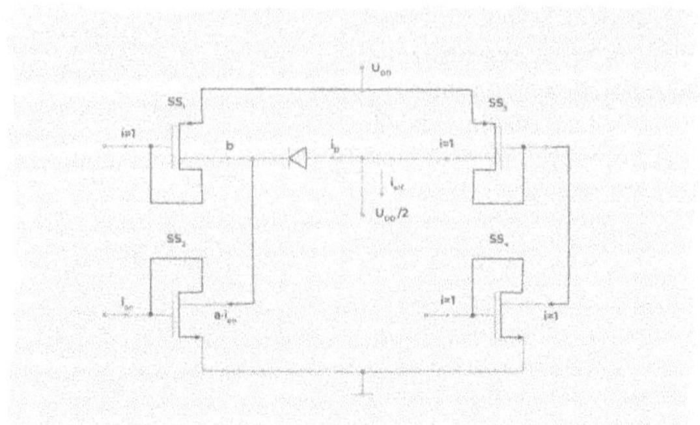

derived from Fig. 10.4a). To achieve better current source characteristics, the current
mirrors can be implemented using cascode technology.

By combining the straight-line segments appropriately, triangular and trapezoidal
functions can also be realized with (10.6) and (10.7).

10.3   Operators and Defuzzification

This chapter describes analog circuit designs for some of the linking operators
presented in Sect. 2.3. They can be derived for current-controlled logic from current
mirrors and nodes. To simplify the derivation, a new operator for linking two fuzzy
sets **A** and **B** is first introduced.

Definition 10.1 (Bounded Difference) Let there be two fuzzy sets $\mathbf{A}, \mathbf{B} \in \mathcal{P}(X)$.
Then the bounded difference of **A** and **B**, written $\mathbf{A}_{-b}\, \mathbf{B}$, is the fuzzy set **C** for which
the following applies:

$$\mu_C(x) = \max[0, \mu_A(x) - \mu_B(x)] \qquad \forall x \in X.$$

10.3.1   Implementation of the Bounded Difference

The circuit shown in Fig. 10.6 with n-MOS current mirrors is used to implement the
bounded difference, which generates a normalized output current i_C as a function
of the input currents i_A and i_B as follows:

$$i_C = \begin{cases} i_A - i_B, & \text{for } i_A \geq i_B, \\ 0, & \text{otherwise.} \end{cases} \tag{10.8}$$

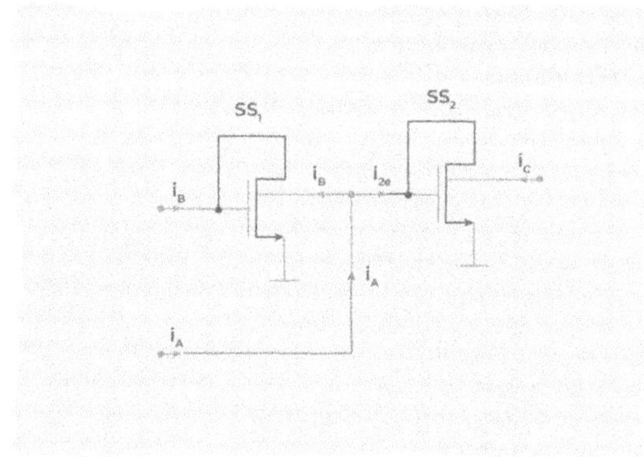

Fig. 10.6 Realization of the bounded difference with n-MOS current mirrors

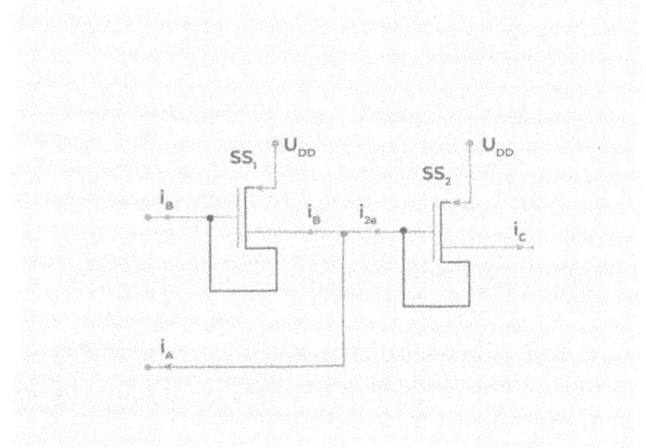

Fig. 10.7 Implementation of the bounded difference with p-MOS current mirrors

In the first current mirror SS_1, the current direction of i_B is first reversed. The node equation then provides the following for the input current i_{2e} of the current mirror SS_2:

$$i_{2e} = i_A - i_B.\qquad(10.9)$$

According to the explanations in Sect. 10.1, the current mirror SS_2 also makes the decision:

$$i_C = \begin{cases} i_{2e}, & \text{for } i_A \geq i_B, \\ 0, & \text{otherwise.} \end{cases}\qquad(10.10)$$

A comparison of (10.9) and (10.10) directly confirms requirement (10.8).

The corresponding p-MOS circuit (shown in Fig. 10.7) is created by replacing the n-MOSFETs with p-MOSFETs and swapping the operating voltages U_{SS} and U_{DD}, whereby the current arrows must be reversed.

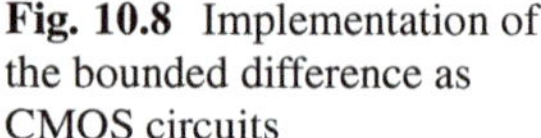

Fig. 10.8 Implementation of the bounded difference as CMOS circuits

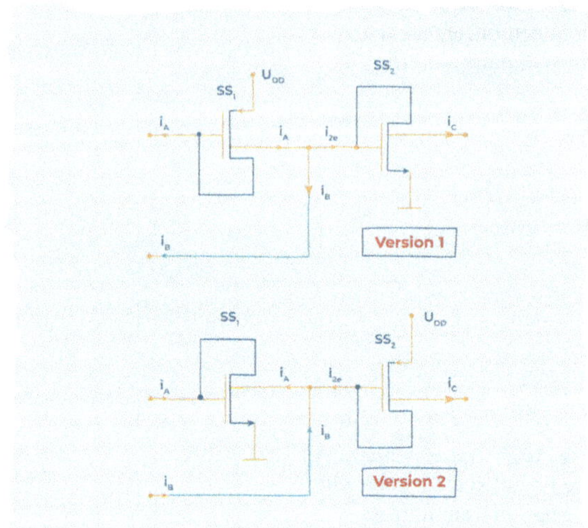

So while positive input and output currents flow into the circuit in the n-MOS version, they flow out in the p-MOS version.

CMOS[3] circuits for the bounded difference can be created by swapping one of the two current mirrors; either the n-MOS current mirror SS$_1$ (in Fig. 10.6) is replaced by a p-MOS current mirror (version 1), or the p-MOS current mirror SS$_1$ (in Fig. 10.7) is replaced by an n-MOS current mirror (version 2). Both versions are shown in Fig. 10.8.

In version 1, positive input currents flow out of the circuit, while a positive output current flows into the circuit. Since the current directions are reversed in version 2, all relevant current direction variations can be realized with the four circuits presented without additional current mirrors. This provides the desired flexibility in the design of complex operator circuits, in which output currents in turn serve as input signals for downstream operators. All of the operator circuits shown in the following sections can be derived from the bounded difference and implemented in the four versions presented. In principle, all available CMOS technologies can be used to manufacture these circuits. In contrast to digital CMOS circuits, however, noticeable operating currents flow here, resulting in static power dissipation. Since the information is mapped directly as a physical signal parameter without encoding, the circuits described can be classified as analog circuits.

---

[3] A complementary metal oxide semiconductor (CMOS) is a MOSFET that uses complementary and symmetrical pairs of p-type and n-type MOSFETs for logic functions.

Fig. 10.9 Implementation of
the complement with n-MOS
current mirror

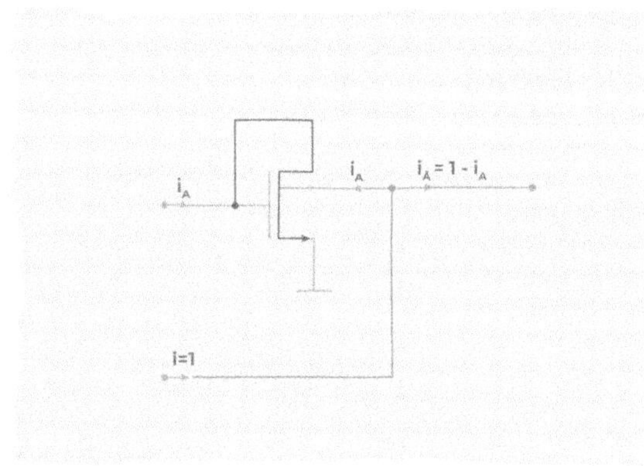

10.3.2 Implementation of the Complement

An n-MOS circuit is derived from the bounded difference by swapping the input currents i_A and i_B in Fig. 10.6 and then setting $i_B = 1$. If no reversal of the output current is desired, the current mirror SS_2 can be omitted. The output current is then:

$$i_{AC} = i_{2e} = 1 - i_A. \tag{10.11}$$

If i_A represents a current membership value of the fuzzy set **A**, then i_{AC} represents the corresponding membership value of the complement $\mathbf{A}^c$ according to (10.11). This results in the circuit shown in Fig. 10.9.

p-MOS and CMOS versions can also be derived using the same principle.

10.3.3 Implementation of the Intersection Operator

The membership values $\mu_C(x)$ of the intersection $\mathbf{C} = \mathbf{A} \cap \mathbf{B}$ are calculated according to Sect. 2.3 using the formula $\mu_C(x) = \min[\mu_A(x), \mu_B(x)]$. This relationship can also be represented as follows:

$$\mu_C(x) = \begin{cases} \mu_A(x) - \big(\mu_A(x) - \mu_B(x)\big), & \text{for } \mu_A(x) \geq \mu_B(x), \\ \mu_A(x), & \text{otherwise.} \end{cases} \tag{10.12}$$

Equation (10.12) can be rewritten as:

$$\mu_C(x) = \mu_A(x) - \mu_{A-_bB}(x). \tag{10.13}$$

According to this, a CMOS circuit for forming the intersection can be derived from version 1 of the CMOS circuit for the bounded difference shown in Fig. 10.8,

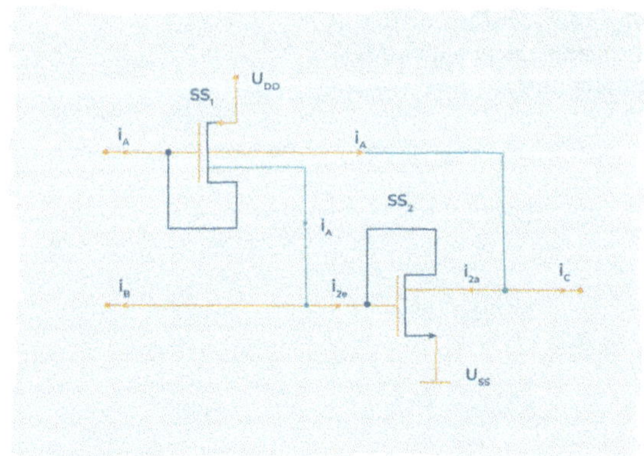

Fig. 10.10 Implementation of the intersection operator in CMOS technology

for example. For this purpose, the positive direction of the output current must be inverted, and a mirrored current i_A must be fed into the output node. This results in a circuit version as shown in Fig. 10.10.

The functionality can be easily verified using the node rule. The following applies:

$$i_{2e} = i_A - i_B \qquad \text{and} \qquad i_C = i_A - i_{2a}. \tag{10.14}$$

Claim 1 ($i_A \geq i_B$) From (10.14) it follows that $i_{2e} \geq 0$, so that the current mirror SS_2 does not block. This means that:

$$i_{2a} = i_{2e} = i_A - i_B,$$
$$\text{therefore} \quad i_C = i_A - i_{2a} = i_B. \tag{10.15}$$

Claim 2 ($i_A < i_B$) From (10.14) it follows that $i_{2e} < 0$, so that the current mirror SS_2 blocks. This means that:

$$i_{2a} = 0,$$
$$\text{therefore} \quad i_C = i_A - i_{2a} = i_A. \tag{10.16}$$

The circuit shown in Fig. 10.10 thus implements the intersection operator.

10.3.4 Implementation of the Union Operator

The membership values $\mu_C(x)$ of the union $\mathbf{C} = \mathbf{A} \cup \mathbf{B}$ are calculated according to Sect. 2.3 using the formula $\mu_C(x) = \max[\mu_A(x), \mu_B(x)]$. This relationship can also be expressed as follows:

Fig. 10.11 Implementation of the union operator in n-MOS technology

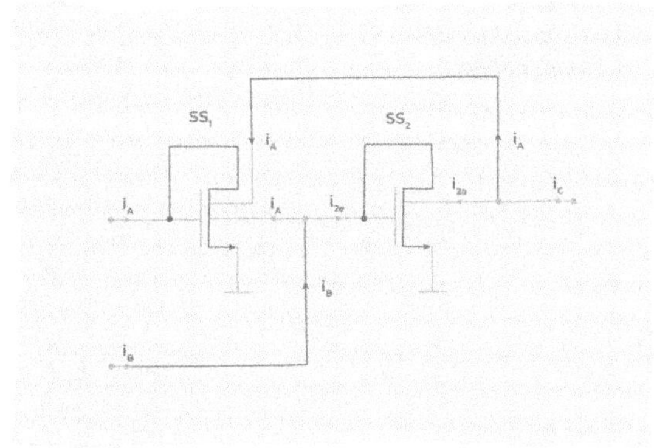

$$\mu_C(x) = \begin{cases} \mu_B(x) + \big(\mu_A(x) - \mu_B(x)\big), & \text{for } \mu_A(x) \geq \mu_B(x), \\ \mu_B(x), & \text{otherwise.} \end{cases} \tag{10.17}$$

Equation (10.17) can be rewritten as:

$$\mu_C(x) = \mu_B(x) + \mu_{A-_bB}(x). \tag{10.18}$$

According to this, an n-MOS circuit for the bounded difference according to Fig. 10.6 can be derived to form the union. For this purpose, the input currents i_A and i_B are first swapped, and then the mirrored current i_A is added to the output current i_C. This results in a circuit version (see Fig. 10.11).

The functionality can be easily verified using the node equations.

10.3.5   Implementation of the Restricted Product

The membership values $\mu_C(x)$ of the bounded product $\mathbf{C} = \mathbf{A} \sqcap \mathbf{B}$ are calculated according to Sect. 2.3 using the formula:

$$\mu_C(x) = \max[0, \mu_A(x) + \mu_B(x) - 1]. \tag{10.19}$$

According to this, a CMOS circuit for forming the bounded product can be derived from the CMOS circuit (version 2) for the bounded difference (according to Fig. 10.8), for example. For this purpose, the input current i_A (in Fig. 10.8) is replaced by the sum $i_A + i_B$, and the input current i_B is replaced by the current $i = 1$. This results in a circuit version (according to Fig. 10.12), which can be easily verified by the node equations.

Fig. 10.12 Implementation
of the bounded product in
CMOS technology

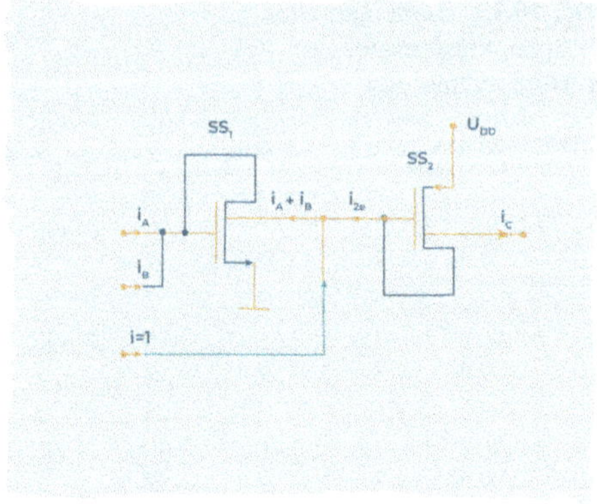

Fig. 10.13 Implementation
of the bounded sum in CMOS
technology

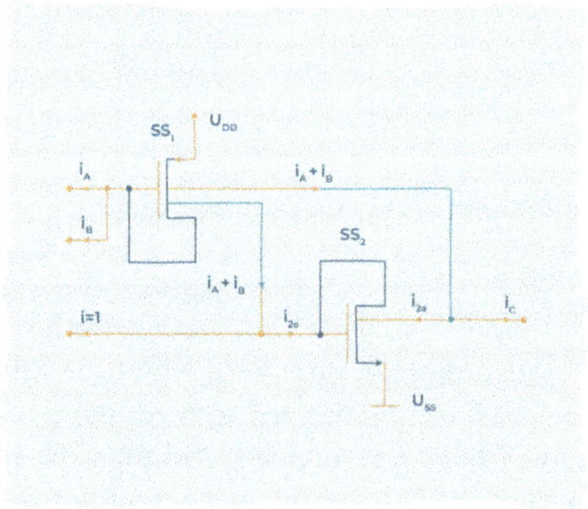

10.3.6   Implementation of the Bounded Sum

The membership values $\mu_C(x)$ of the bounded sum $\mathbf{C} = \mathbf{A} \sqcup \mathbf{B}$ are calculated (according to Sect. 2.3) using the formula:

$$\mu_C(x) = \min[1, \mu_A(x) + \mu_B(x)]. \tag{10.20}$$

According to (10.20), a CMOS circuit for forming the bounded sum can be derived from the circuit for the intersection operator (see Fig. 10.10). To do this, the input current is replaced by the sum $i_A + i_B$, and the second input is replaced by current 1. This results in a circuit version as shown in Fig. 10.13.

The functionality of the circuit can again be easily verified using the node equations.

10.3.7 *Implementation of Defuzzification*

A circuit proposal according to Kettner et al. (1992) for the defuzzification of the output variables is based on a modified centroid method for calculating the centroid abscissa value y_s as the control variable according to (8.1), which is calculated as:

$$y_s \approx \frac{\sum_{i=1,\dots,n} \mu_i \, \ddot{y}_i}{\sum_{i=1,\dots,n} \mu_i}, \qquad (10.21)$$

where $\ddot{y}_i$ represents the centroid abscissa value of the complete membership function, which represents the consequence term of rule (i), while μ_i indicates the degree of membership at the point $\ddot{y}_i$ (see Example 10.1). The evaluation of (10.21) can also be interpreted as meaning that the membership functions of the consequence terms are given by singletons. To calculate the value, a total of n multiplications, $2(n-1)$ summations, and one division must be performed.

Analog multipliers based on differential amplifier stages can be used for the multiplications in the numerator polynomial of (10.21); the current sources are controlled by the multiplier via current mirrors. If both the values of the singletons $\ddot{y}_i$ and the membership values $\mu(\ddot{y}_i)$ are represented by normalized currents, the summations can be performed in circuit nodes. The reference variable for the singletons is the variability range of the control variable.

The schematic diagram of a current-controlled divider is shown in Fig. 10.14. It is constructed from n-MOS, p-MOS, and bipolar transistors and can be implemented using BiCMOS technology.[4] Its task is to divide the counter current i_z by the denominator current i_N. The result is the normalized output current $i_{z/N}$, which represents the control variable y_s.

The circuit consists of two differential amplifier stages with bipolar NPN transistors and three CMOS current mirrors. In the first differential amplifier stage, the mirrored denominator current i_N is split into two partial currents, one of which is also specified by the mirrored counter current i_z. The circuit developer must therefore ensure that, in accordance with (10.21), $i_N \geq i_z$ applies to the impressed currents. The current ratio i_z/i_N results in a clear differential voltage at the base electrodes. This is transferred to a second differential amplifier stage and divides the current $i = 1$ in the same current ratio i_z/i_N. This results in the following transfer equation for the circuit:

$$i_{z/N} = i_z/i_N. \qquad (10.22)$$

Example 10.1 A simplified version of (10.21) is obtained when the membership function of the fuzzy output variable is sampled at predefined points in the basic

---

[4] Bipolar-CMOS technology merges bipolar junction transistors with CMOS on the same chip, making it ideal for signal processing, offering high performance, efficiency, and integration.

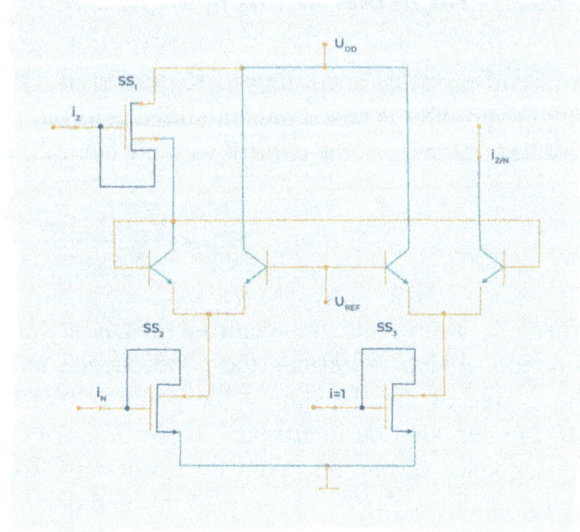

Fig. 10.14 Schematic diagram of a current-controlled divider in BiCMOS technology (according to Kettner et al., 1992)

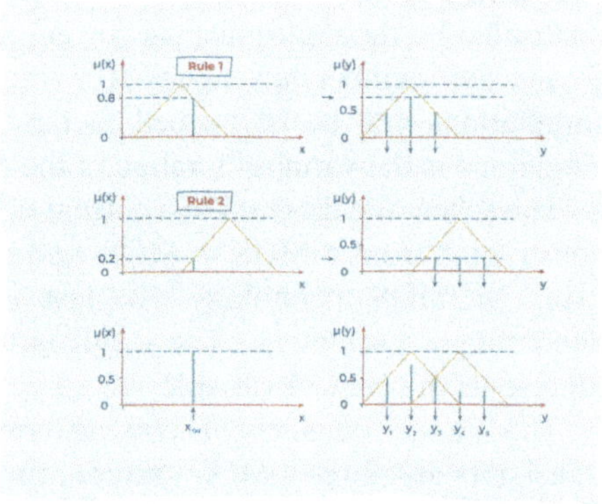

Fig. 10.15 Control value calculation with simplified defuzzification

range. In current-controlled circuit technology, multiplication can then also be performed with variable programmability using appropriately set mirroring factors. Furthermore, the calculation of the membership values for the output variable can be reduced to specified points in the basic range. Constant current sources can be used for this purpose. When applying the max-min inference method and combining the individual conclusions, the calculation schema (shown in Fig. 10.15) is obtained.

A controller is shown with one input variable and one output variable, whose transfer behavior results from two production rules. Fuzzification provides the activation degrees $\beta_1 = 0.8$ for rule 1 and $\beta_2 = 0.2$ for rule 2. For both the two sub-conclusions and the conclusion, membership values are only determined at the five points $y_1, \ldots, y_5$. This results in the control value y_s being calculated as:

$$y_s = \frac{y_1 \mu(y_1) + \cdots + y_5 \mu(y_5)}{\mu(y_1) + \cdots + \mu(y_5)}. \tag{10.23}$$

Fig. 10.16 Block diagram of
a simple fuzzy controller

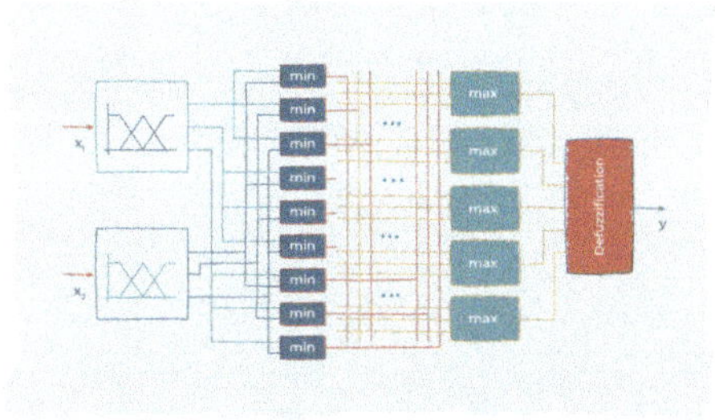

10.4 Overall Structure

With the components described in the previous chapters, a simple fuzzy controller
can now be constructed. A block diagram for two input variables, one output
variable, and 15 rules is shown in Fig. 10.16.

First, the input variables x_1 and x_2 are fuzzified using three membership functions
each. In the condition part of the rules, only AND operations are permitted, which
are represented by intersection operators. The rules are established in a composition
matrix by means of conductive connections. The five maximum operators represent
the fixed points $\ddot{y}_i$ for calculating the sharp control value according to (10.23) and
can each take up to three rules into account. They select the largest membership
values $\mu(\ddot{y}_i)$ at the points $\ddot{y}_i$ and forward them to the defuzzifier.

References

Bothe, H.-H. (1995). *Bewertung mit unscharfen Mengen*. Technische Hochschule Ilmenau.

Fattaruso, J. W., & Meyer, R. G. (1987). Fuzzy control strategies for industrial applications. *Fuzzy Sets and Systems, 24*(1), 67–78.

Heite, R., Bothe, H.-H., & Zimmermann, H.-J. (1989). Fuzzy control: A survey. *Fuzzy Sets and Systems, 30*, 1–28.

Kettner, T., et al. (1992). Fuzzy control of complex dynamic systems. *Control Engineering Practice, 1*(4), 673–681.

Watanabe, T., Hirota, K., & Sugeno, M. (1990). A design of fuzzy controllers. *Fuzzy Sets and Systems, 30*, 1–28.

Yamakawa, T., & Miki, T. (1986). Application of fuzzy logic to industrial process control. *Fuzzy Sets and Systems, 18*(2), 155–165.

Zhijian, J., & Hong, Z.-J. (1991). Design of fuzzy controllers using linguistic rules. *Fuzzy Sets and Systems, 42*(1), 45–54.

Chapter 11
Toward Computing with Words

Abstract This chapter is about the paradigm shift in computer science, when artificial intelligence starts to process information using words and propositions from natural language. It aims to build machines that reason like humans by handling the inherent vagueness and imprecision of human language holistically.

This final chapter is about why our systems need to move beyond numbers and start reasoning in natural language. It integrates all chapters from Bothe (1995) and expands them with computing with words. We once again revisit all the individual elements presented by Bothe (1995) and integrate them into Zadeh's (2012) most recent framework of computing with words.[1]

What does it mean to describe someone as "mostly honest" or a day as "kind of cold"? To a human, these phrases make perfect sense. To a computer, they traditionally mean nothing at all. Yet these are exactly the kinds of expressions we use to make decisions—how much to trust, at when to act, and whether to wait. They are fuzzy, informal, and rooted in language, not numbers. And they present a deep challenge to traditional computing. That is why we needed a new way forward. In equating fuzzy logic with *computing with words*, Zadeh (1996) proposed an idea as simple as it was radical. Instead of reducing human language to numbers before computation can begin, we should find a way to compute with the language itself.

Definition 11.1 (Computing with Words) Computing with words is not a variant of natural language processing, nor is it a soft version of classical logic. It is a formal framework for representing and reasoning with information that is *imprecise by nature* (i.e., statements like "usually the train arrives on time" or "it is highly unlikely the price will fall"). Computing with words offers a way to handle such information without forcing it into the sharp corners of binary logic.

This matters because so much of our decision-making is based on *perception*. Whether we are evaluating a risk, interpreting a social signal, or judging the mood of

---

[1] Zadeh's (2012) presentation, which has been published in a book form, is highly recommended and is still far too little known in the AI world.

a room, we rely on language to express uncertainty. This opens the door to systems that can model that uncertainty, not as error or noise, but as a fundamental part of intelligent reasoning.

This chapter concludes with the foundations and methodology of computing with words (i.e., also known as future development of fuzzy logic). Section 11.1 introduces the core ideas and their motivation. Section 11.2 explores how meaning is "precipitated" (i.e., transformed into a form that machines can reason with) and how computation follows from that. Section 11.3 offers illustrative examples that show how an implementation with fuzzy cognitive mapping might work in practice.

In a world increasingly shaped by systems that must interact with humans on *human terms*, in all their frustrating imprecision, computing with words is not a novel approach but a necessity.

11.1 Precisiation of Meaning

Zadeh's vision for computing goes beyond numbers, and it begins with language:

Example 11.1 (Challenges) Most people do not speak in equations. Ask someone how long their commute takes, and you'll hear something like "Usually about 45 minutes." We instinctively hedge our estimates with words like "about," "around," "usually," or "more or less." For us, this is normal, but it presents a problem for computers trained on hard data and binary logic.

Zadeh (2012) saw this gap as more than just an inconvenience; it is a fundamental mismatch between human thinking and machine reasoning. In response, he introduced a deceptively simple idea: *computing with words*. Instead of replacing numbers, it would expand computation to include the kinds of things people actually say.

Zadeh (1965) had already developed fuzzy logic, the introduced framework that allows systems to reason with degrees of truth, where a statement can be 80% true and 20% false. Computing with words takes that framework further. It asks, what if the inputs to a system were not just numbers, but words, phrases, and perceptions? And what if we treated those words not as noise, but as data?

Example 11.2 (Reasoning) To see how computing with words works, consider two statements:

- Dana is 25 years old.
- Tandy is a few years older than Dana.

In traditional computing, this becomes:

$$\text{Tandy} = \text{Dana}\,(25) + 3 = 28.$$

In computing with words, both "young" and "a few" are considered fuzzy numbers (i.e., labels with flexible boundaries as introduced in Chap. 4). Instead of using precise values, the system applies *fuzzy arithmetic*, combining the uncertainty in "young" with that in "a few" to estimate Tandy's age. The outcome is not a specific number but a *possibility distribution* (i.e., a soft-edged, calculated guess).

This is the essence of computing with words: It manipulates *meaning*, not just measurements.

11.1.1 A System Built on Perception

Computing with words does not try to sanitize or reduce language. It does the opposite; it *preserves imprecision* and treats it as information. That is a major departure from classical computation, where every variable must be crisply defined, and every answer must be exact.

Most of the world does not operate that way. We say things like "most Swedes are tall" or "it is very unlikely we will arrive on time." These are judgments, not data points. Still, we use computational problems expressed in *natural language* to reason, plan, and decide. Traditionally, such problems are seen as ill-posed. Computing with words asks, what if they are simply posed differently?

Consider the question: "What is Vera's age?" We might find out that Vera has a daughter in her mid-30s and that women usually give birth between the ages of 20 and 40. That is vague, but still helpful. Computing with words views each piece of information as a restriction (i.e., a fuzzy constraint on what Vera's age could reasonably be). The answer, therefore, is not a single number but a fuzzy estimate shaped by overlapping linguistic clues. This is what Zadeh called *information as restriction*. And it forms the backbone of computing with words' approach to reasoning.

11.1.2 A Comparison with Natural Language Processing

It is tempting to assume that computing with words is just another version of natural language processing—it is not. Natural language processing is about understanding, parsing, and generating human language. Computing with words, on the other hand, is about reasoning with it. The difference is subtle but important.

Natural language processing might help a system recognize that "a few years older" refers to age. Computing with words lets the system *reason* with that information (see Sect. 11.2) without first converting it into a hard number. That is the key distinction; computing with words does not extract language to discard it. It *stays within language* as long as possible, working with phrases like "probably," "usually," or "somewhat likely," treating them as labels of fuzzy sets (Chap. 5).

Instead of forcing them into a binary form, it allows them to keep their natural uncertainty.

Example 11.3 Computing with words begins with a question q, such as "What is the average height of Germans?" Then comes the *information set* I, which might include "Most Germans are tall." From there, the system constructs an answer:

$$\text{ans}(q \mid I).$$

To compute that answer, the system must specify the components. "Tall," for instance, becomes a fuzzy set defined over the space of possible heights. "Most" becomes a fuzzy quantifier. Together, they impose a restriction: The proportion of tall Germans is high. That restriction informs a downstream computation, like average height, using tools like the *extension principle.*

This approach allows computing with words to operate in domains where statistical models struggle. You don't need massive datasets. You need meaningful input, even if it is vague, or especially if it is vague.

11.1.3   Computing with Words Levels

Most current applications of the presented fuzzy logic operate at what Zadeh (2012) called *level 1*. These include systems like washing machines, autofocus cameras, or thermostats. They use fuzzy rules to control output: "If temperature is high, set fan to fast." Inputs and outputs are labeled in words, but the rules are fairly simple.

Computing with words *level 2*, which still needs to be (further) developed, deals with more complex language. Not just "hot" or "cold," but statements like "it is very unlikely the situation will improve." These statements include context, expectation, and world knowledge. They reflect how people actually speak when they are uncertain . Zadeh (2012) believes this is where the true power of computing with words lies.

The difference between level 1 and level 2 is like the difference between reflex and reflection. Level 1 systems respond to conditions, and level 2 systems *reason through them.*

11.1.4   Reasons to Compute with Words

There are at least three reasons why we compute with words: first, *necessity* (i.e., sometimes words are all we have); second, *advantage* (i.e., words are often cheaper to elicit than numbers, especially when asking people); and third, *expedience* (i.e., words summarize complex impressions efficiently). In other words, they are not just a fallback; they are an upgrade.

Example 11.4 Think of a doctor's note that reads "Patient seems tired and slightly disoriented" or a weather forecast "mostly sunny with a chance of showers." These are not simple placeholders for numbers; they are useful as-is because they reflect how humans *perceive* the world.

By building systems that can compute with such input, we open the door to more natural interactions (i.e., the ones based not on data entry, but on dialogue). In fields ranging from medicine to trust modeling to everyday AI, that is a critical shift.

Of course, to hold a conversation, a system first needs to understand what you're saying. It needs to *cross from perception to computation*. That is where things start to get interesting.

11.2   Computation with Meaning

To compute with language, a system must first make meaning precise. This chapter explains how that precision is achieved and how fuzzy restrictions allow reasoning to begin.

11.2.1   *Precisiation and the Challenge of Linguistic Meaning*

Human language resists precision. It is filled with approximations, flexible terms, and meanings that change with context. For computing systems based on binary logic, this presents a challenge: How do you process an input like "often" or "quite high" when your framework requires either 0 or 1?

Example 11.5 Our response should not be to oppose this ambiguity but to formalize it. We propose the idea of *precisiation*, a process for translating natural language statements into structures that could be handled mathematically, yet without forcing them into rigid categories. Importantly, this did not mean simplifying or discarding what people say. It means maintaining the intent and nuance of a statement while making it computationally meaningful.

At the core of this idea is the *meaning postulate*, which suggests that a proposition does not merely state a fact; instead, it restricts a variable. When someone says, "Robert is very honest," they are not assigning a *numerical* value to a trait; they are imposing a constraint, albeit a soft one, on the range of "honesty scores" Robert might *plausibly fall within*. This restriction forms the foundation for how the system interprets and reasons with the statement.

However the strength of this approach does not just lie in what it allows a system to interpret; it lies in what it makes possible next.

11.2.2   Possibility, Probability, and Z-Number Restrictions

To model vague statements, the system needs to adopt a set of constraints or restrictions. These vary in form depending on the type of information involved.

The most fundamental of these is the *possibilistic* restriction. It describes uncertainty through degrees of membership in a fuzzy set. If "tall" is depicted as a smooth curve over the range of human heights, then a person who is 185 cm tall might belong to the set "tall" to a degree of 0.8. Crucially, these are not strict cutoffs, but adaptable contours.

Then there are *probabilistic* restrictions that manage uncertainty through *chance*. These are helpful when working with statistical expectations, like "there is a 70% chance of rain tomorrow." Although probabilistic models are common in traditional computing, they lack the nuance to handle *graded linguistic terms*. They express distributions, not perceptions.

The concept of *Z-numbers* bridges these worlds. Z-numbers encode not only the fuzzy estimate (e.g., "about 25") but also a fuzzy reliability rating (e.g., "usually"). In essence, they describe a value and how confidently we hold that value, within the same construct (Zadeh (2011)).

While different situations demand various kinds of restrictions, what links them is their capacity to translate *nuanced information* into *precise reasoning*. And to achieve this, they require a solid foundation.

11.2.3   The Explanatory Database and the Precisiated Canonical Form

Language is never interpreted in a vacuum. This is as true for humans as it is for AI systems. To understand a statement like "most Germans are tall," a system needs access to background knowledge: Who qualifies as Swedish, what counts as "tall," and how "most" is defined. We formalize this requirement through the notion of an *explanatory database*.

Definition 11.2 (Explanatory Database) An explanatory database consists of a set of auxiliary relations that define the variable and its restriction. For the example above, this might include a population-height dataset, a fuzzy set defining "tall," and a fuzzy quantifier that captures what "most" means. These elements enable the system to interpret the original sentence computationally.

The result is a *precisiated canonical form*:

$$X^* \ \text{is}_r \ R^*.$$

Here, both the variable X and the restriction R have been fully specified, ready for further computation. However, precisiation goes a step further; not just to tidy up language, but to give it structure on how perception is represented, opening the way for generalization and embracing variables.

11.2.4 Linguistic Variables and Granulation

As we have seen, a key innovation is the idea of the linguistic variable (Zadeh, 1975). Instead of assigning specific numerical values like 17.4 or 62.1, a linguistic variable might have values such as "young," "middle-aged," or "old" (see Chap. 4). These values are defined as fuzzy sets and often overlap. This overlap enables systems to operate similarly to humans, with subtle distinctions and gray areas. This is where *granulation* comes in.

Definition 11.3 (Granulation) Granulation means dividing a domain into understandable, overlapping categories, or granules, that mirror human judgment. For example, if someone is 35, they might fall into the "young" category at 0.6 and the "middle-aged" category at 0.4. There is no need to assign a strict label.

Granulation allows machines to recognize the same flexible boundaries that humans use when they think, speak, or decide. It also offers the framework for fuzzy inference. And to reason reliably across many such fuzzy structures, a system benefits from recognizing patterns in how they're formed.

11.2.5 Protoforms' General Reasoning Templates

Propositions like "most Germans are tall" or "many meetings were unproductive" may differ in content but share a logical structure. Zadeh introduced the idea of *protoforms* to capture this.

Definition 11.4 (Protoform) A protoform is an abstract template that generalizes the structure of a precisiated proposition. "Most As are Bs" is a protoform. Specific instances apply to A and B. This abstraction enables a system to use learned patterns or reasoning strategies across different domains without starting from scratch each time (Zadeh (2006)).

Protoforms should not be viewed merely as a technical convenience. Instead, they embody the fundamental *regularities* of human reasoning through language. They acknowledge that, although human reasoning may appear unstructured at first glance, it ultimately reveals consistent patterns. In computing with words, these patterns serve as the blueprints for extending inference beyond the limited scope of a single example.

It is when a proposition has been precisiated and mapped to a protoform that it can take the next step: from interpretation to *computation*.

Fig. 11.1 Basic structure of computing-with-words-based model of computation

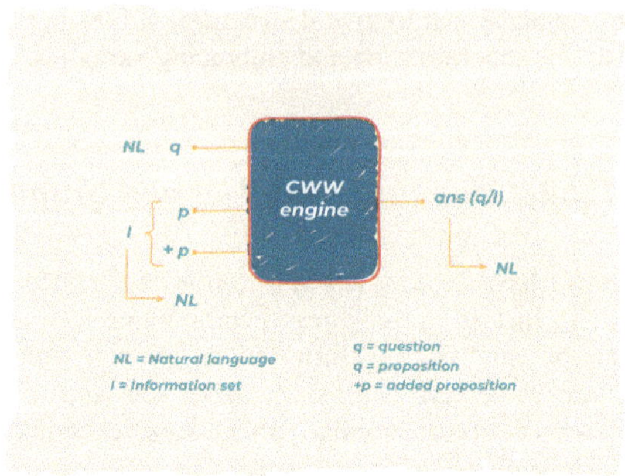

11.2.6   Computation with Restrictions

Once language has been made precise, the system is prepared to reason with it, but instead of employing classical logic or arithmetic, it involves computing over *restrictions*.

Example 11.6 (Average Heights) A question is asked (e.g., "what is the average height of Germans?"). A set of propositions is given, such as "most Germans are tall." Each proposition is clarified into a fuzzy restriction. These constraints are then combined and propagated to determine the answer (see Fig. 11.1).

This process demonstrates how people *reason with partial knowledge*. If we know that most individuals in a group are tall, we might reasonably assume that the average height exceeds a certain threshold. All that computing with words does is formalize that intuition, and to do this, it depends on one of the most powerful principles in fuzzy logic.

11.2.7   The Extension Principle and Fuzzy Functions

The extension principle explains how a function maps fuzzy inputs to fuzzy outputs. If a function takes variables X_1 through X_n and produces Y, and each X_i is defined by a fuzzy set, the principle specifies how the fuzzy set for $\mathbf{Y}$ is determined based on the contributions of input values to each possible output.

The principle can be expressed as:

$$\mu_Y(v) = \sup\left\{\min\left[\mu_{X_1}(x_1), \ldots, \mu_{X_n}(x_n)\right] \mid f(x_1, \ldots, x_n) = v\right\}.$$

This formula enables a system to carry the ambiguity of inputs into its conclusions. There is no reduction to a single clear output unless one is explicitly required.

It is this propagation of uncertainty that allows computing with words to operate in real-world environments, where complete data is scarce and imprecision is common, not unusual. And where data is limited but structured knowledge is present, it reveals unexpected opportunities.

Example 11.7 (Estimating an Average) Let us revisit the problem of estimating the average height of Germans. The only information available is "most Germans are tall." There is no comprehensive dataset. Instead, the only "data" is a linguistic statement.

To proceed, the system defines "tall" as a fuzzy set over the height domain. "Most" becomes a fuzzy proportion. The proposition is interpreted as a *weighted constraint*: A significant portion of the population has a high degree of membership in "tall." From there, possible average heights are calculated based on assumed or sampled height distributions, adjusted by their degrees of membership and the "most" quantifier.

The result is not a single figure, but a fuzzy *estimate* (i.e., a probability distribution over plausible averages). The answer matches the nature of the question: imprecise yet informative.

This example illustrates how language, when correctly structured, can act as a substitute for data. And this method is not restricted to averages.

11.2.8 Generalizing Inference

As we have seen, in many fuzzy control systems, inference is performed through rule sets: *IF X is A THEN Y is B*. Such rules are valuable in systems like thermostats or cruise control, where inputs and outputs are closely linked.

Computing with words generalizes this. It does not require propositions to fit into IF-THEN form. It supports chains of reasoning constructed from arbitrary propositions (i.e., possibly involving quantifiers, adverbs, or probabilistic overlays). In doing so, it broadens the scope of fuzzy systems from control logic to interpretive judgment.

This capacity becomes essential in domains where the input is less about quantifiable frameworks (e.g., temperature) and more about intangible factors (e.g., trust, mood, or risk).

11.2.9 Lessons Learned

Outlined in Fig. 11.2, computing with words occurs in two closely connected stages: First is precisiation, where the vagueness of natural language is shaped

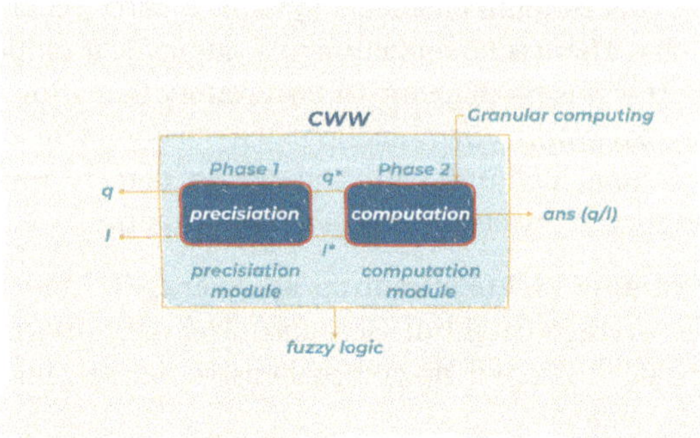

Fig. 11.2 Precisiation and computation employ the introduced machinery of fuzzy logic

into a structured form (i.e., a space of fuzzy variables, constraints, and quantified relations). Then computation is where that structure is no longer only interpreted but also reasoned with. This transition reflects how humans think, from understanding meaning to drawing conclusions.

This two-phase process allows systems to handle information that defies clear boundaries (i.e., statements that are common, likely, or vaguely descriptive). It challenges the binary thinking found in traditional computation. It does not aim to simplify what we say. Instead, it seeks to understand it as it is.

That shift from what we say to what we compute is subtle but essential. It paves the way for machines to interact with human input in a manner that more closely resembles its natural appearance. It also changes what we consider "data."

And while the logic behind this transformation is compelling, it is the practical results that reveal its true strength. Because only when this approach is applied (i.e., when vague statements meet real-world reasoning as we present next to close this chapter) do its implications become fully apparent.

11.3   Mapping Meaning

How do fuzzy cognitive maps bridge human reasoning and machine interpretation? As a straightforward implementation of the presented computing with words paradigm, we are closing up the book here.

11.3.1   From Restrictions to Reasoning

In the previous chapter, we explored how fuzzy restrictions, once formally defined, can be used to compute with meaning. A system could handle a single vague statement (e.g., "most Germans are tall") and derive a soft, reasoned estimate of the average height, but that is just one proposition.

In reality, human reasoning rarely works in isolation. It weaves together many such propositions (i.e., causes, effects, expectations, and counterfactuals) into a seemingly tangled and incalculable web of impressions and influence.

To capture and *model* that kind of interconnected thinking, we need more than a simple rule. We need a framework, and that framework is the fuzzy cognitive map.

First developed as a tool to simulate human decision-making, fuzzy cognitive maps are a powerful addition to our computing with words toolbox. They provide a way to model not only how concepts are *perceived* but also how those perceptions *influence* each other or how one fuzzy impression *flows into* another, *alters* its shape, or *amplifies* its strength. In short, fuzzy cognitive maps do not just process imprecise concepts; they also *structure* them.

11.3.2 Fuzzy Cognitive Maps

The roots of fuzzy cognitive maps trace back to cognitive maps, which modeled how people perceive and communicate causal relationships in complex systems (Axelrod, 1976). These early maps were directed graphs (i.e., concepts connected by arrows indicating influence). However, the structure was binary and qualitative in nature.

Kosko (1986) extended this idea by introducing fuzziness into both concepts and their connections. The result was the fuzzy cognitive map, a system of nodes representing concepts, with degrees of activation, and weighted edges capturing the *strength* and direction of causal influence (Fig. 11.3).

Definition 11.5 (Fuzzy Cognitive Maps) Formally, a fuzzy cognitive map consists of:

- A set of concepts $C = \{C_1, C_2, \ldots, C_n\}$, each with a state value $C_i \in [0, 1]$
- A connection matrix $W = [w_{ij}]$, where each weight $w_{ij} \in [-1, 1]$ represents the influence of concept C_j on concept C_i
- An inference mechanism based on iterative updates

The fuzziness appears in two ways:

1. The *state values* of concepts reflect partial activation or presence.
2. The *weights* between concepts reflect varying causal intensities, from strongly negative (inhibitory) to strongly positive (reinforcing).

This structure enables fuzzy cognitive maps to reflect how humans often think, not in absolutes, but in tendencies and trends.

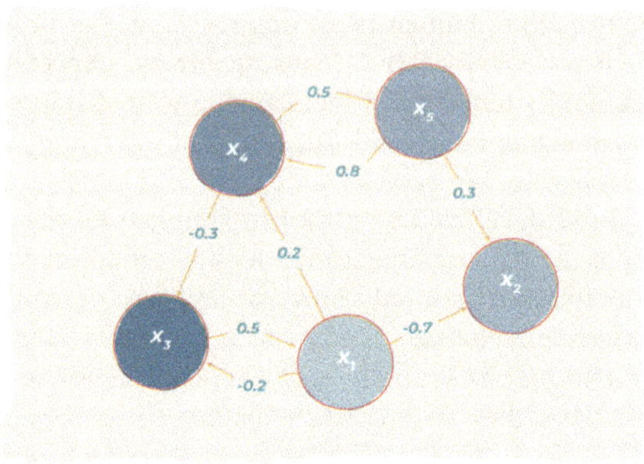

Fig. 11.3 A simple representation of a fuzzy cognitive map with five generic vertices and weighted edges showing the relationships between the set of concepts

11.3.3   Fuzzy Cognitive Mapping

The reasoning process in a fuzzy cognitive map unfolds dynamically. At each time step t, the activation level of each concept C_i is updated based on the current states of other concepts and the weights of their connections:

$$C_i^{(t+1)} = f\left(\sum_{j=1}^{n} C_j^{(t)} \cdot w_{ji}\right)$$

with:

- w_{ji} is the weight of the influence from concept C_j to C_i.
- f is a transfer function (i.e., used to constrain the result to the interval $[0, 1]$).

  Typical choices for f include:

- *Sigmoid functions*, which smooth large inputs and produce soft thresholds
- *Trivalent or bivalent functions*, used when a system requires discrete reasoning (e.g., on/off, agree/disagree, etc.)
- *Hyperbolic tangent or step functions*, which allow sharper cutoffs

The purpose of the transfer function is not only to normalize values but also to control feedback, limiting runaway influence and helping the system settle into realistic equilibria.

As the system iterates, it may converge to a *fixed point* (i.e., all values stabilize), a *limit cycle* (values oscillate in a repeating loop), or exhibit *chaotic dynamics* (unpredictable changes due to feedback loops and nonlinearities). These behaviors are not quirks; they matter. A fixed point might reflect policy consensus; a cycle could represent repeating community behaviors; chaos might signal instability in a system under stress.

Example 11.8 (Public Health) Consider a simple fuzzy cognitive map designed to explore public health in a city. Concepts might include:

- C_1: "Air pollution"
- C_2: "Respiratory illness"
- C_3: "Access to healthcare"
- C_4: "Preventive behavior"

Edges represent influence:

- $w_{21} = 0.7$: Pollution increases. illness
- $w_{23} = 0.6$: Healthcare reduces. illness
- $w_{42} = 0.5$: Illness encourages. preventive action

Initial activations are assigned: high pollution, low preventive behavior, and moderate access to care. Then the system is run. As each concept updates, we observe the ripple effects: Perhaps respiratory illness rises and then stabilizes as preventive behavior increases.

The value of this model lies in its *what-if reasoning*. We can simulate the effect of policy changes, environmental shifts, or social interventions by tweaking a single node. If we increase access to healthcare, how does that impact the incidence of illness? If illness decreases, does preventive behavior drop too?

This scenario-based flexibility is particularly useful in fields such as environmental planning and management. Imagine a second fuzzy cognitive map modeling climate resilience:

- "Green infrastructure," "flooding risk," "public awareness," "policy strictness," and "community health" become the concepts.
- Linguistic rules such as "stricter policies *usually lead* to more green infrastructure" or "more awareness *typically reduces* flooding risk" are encoded in fuzzy weights.

Fuzzy cognitive maps are also capable of *feedback amplification* (e.g., a single change to one node might, through multiple rounds of influence, compound or cancel effects). This echoes how human understanding deepens or shifts. Not all effects are immediate or local.

What results is not just a diagram, but a simulation of conceptual reasoning. In the computing with words paradigm, this enables systems to operate with *soft logic that mimics human cognition.*

11.3.4 Constructing Methods

As with any model, construction matters. The strength of a fuzzy cognitive map lies in how faithfully it reflects expert understanding or empirical patterns. There are three standard methods: first, expert-based; second, data-driven; and, third, hybrid methods.

1. Expert-Based Methods

This approach starts with domain knowledge. Experts:

- Identify key concepts.
- Define directional relationships.
- Assign fuzzy weights based on linguistic judgments.

Linguistic labels (e.g., "strongly increases" or "weakly inhibits") are mapped to fuzzy numbers (e.g., $+0.8$, -0.3). These mappings may be standardized or adapted from membership functions. Aggregation operators (e.g., fuzzy mean or weighted voting) combine inputs from multiple experts. Weights can be "defuzzified" if needed.

Example 11.9 Consider an urban planner developing a fuzzy cognitive map for mobility planning. Interviews with stakeholders reveal concepts like "bike infrastructure," "traffic congestion," and "public transport appeal." Rules such as "more bike lanes *usually reduce* traffic" are added, and fuzzy strengths are agreed upon. As a result of the multiple additions, the map evolves through consensus and iterative refinement.

2. Data-Driven Methods

With time-series data, machine learning methods can estimate the fuzzy cognitive map structure. Algorithms like Hebbian learning,[2] differential evolution, or genetic programming derive both the set of concepts and the connection matrix.

Such models can capture unexpected interactions but may sacrifice interpretability. A model might find that "rainfall intensity" negatively affects "bus punctuality" with a weight of -0.65. The data justifies it, but without expert validation, the result may appear arbitrary or contextless.

Studies by Papageorgiou and Froelich (2017) have demonstrated how evolutionary algorithms can be used to develop adaptive fuzzy cognitive maps that refine themselves as new data becomes available. Similarly, Stylios and Groumpos (2000) demonstrated how such models can be utilized in process control and industrial systems with minimal human intervention.

3. Hybrid Methods

Here, experts define the structure (e.g., nodes and links), while algorithms fine-tune the weights. A public health fuzzy cognitive map might be built from practitioner

---

[2] Hebbian learning can be summarized as "neurons that fire together, wire together." It is a principle where the connection strength between two neurons increases when they are simultaneously active, forming stronger pathways for learning and memory.

insights and then calibrated using hospital records, mobility data, or environmental sensors.

Example 11.10 (Disaster Resilience Planning) Experts identify key concepts: "emergency preparedness," "public trust," "media messaging," "resource allocation," and "community response." Fuzzy causal links are defined: "Messaging *strongly affects* trust," and "Trust improves preparedness." Then, behavioral data from prior events is used to adjust the weights. The result is a nuanced and grounded system that can be used to test future interventions or anticipate potential vulnerabilities.

This approach balances rigor with readability. It preserves the traceability of linguistic reasoning while improving accuracy.

Regardless of the method, there is one overlooked factor that can critically shape a fuzzy cognitive map's power: the choice of concepts. Concepts that are too broad ("well-being") or too narrow ("flu incidence on Tuesday mornings") can weaken interpretability or limit generalization. Concepts should reflect the granularity of human language, being broad enough to be meaningful yet specific enough to be useful.

In the computing with words context, these concepts are often shaped by language itself (e.g., drawn from interviews, documents, or observed discourse). It is important to remember that the strength of a fuzzy cognitive map lies not only in the math behind it but also in the *narrative logic of the concepts it connects*.

11.3.5 A Framework for Fuzzy Reasoning

Throughout this concluding chapter, we have followed a progression (i.e., from vague language to formal restriction, to structured inference). What began as a problem of understanding what people say has become a system for computing with what they mean.

Fuzzy cognitive maps close that loop. They don't simply process words. They capture relationships between ideas: not as binary links, but as weighted influences. They encode how people think through ambiguity, not just how they describe it.

They also reflect a more profound shift in how we design computational systems. Traditional models treat uncertainty as noise, something to be filtered or ignored. Fuzzy cognitive maps, like computing with words itself, treat it as a form of information. They allow systems to accept uncertainty, shape it, and reason with it.

As a model of human-like cognition, fuzzy cognitive maps offer structure without rigidity, inference without absolutes, and logic that speaks the language of experience. They bring the project of computing with words full circle, from what we say to what we compute, to how it all connects.

References

Axelrod, R. (1976). *Structure of decision: The cognitive maps of political elites*. Princeton University Press.

Bothe, H.-H. (1995). *Bewertung mit unscharfen Mengen*. Technische Hochschule Ilmenau.

Kosko, B. (1986). Fuzzy cognitive maps. *International Journal of Man-Machine Studies, 24*(1), 65–75. https://doi.org/10.1016/S0020-7373(86)80040-2.

Papageorgiou, E., & Froelich, W. (2017). Fuzzy cognitive maps for applied sciences and engineering—from fundamentals to extensions and learning algorithms. *IEEE Transactions on Systems, Man, and Cybernetics: Systems, 47*(1), 1–14. https://doi.org/10.1109/TSMC.2016.2562501.

Stylios, C. D., & Groumpos, P. P. (2000). Fuzzy cognitive maps: A soft computing technique for intelligent control. *Fuzzy Sets and Systems, 111*(1), 1–26.

Zadeh, L. A. (1965). Fuzzy sets. *Information and Control, 8*(3), 338–353.

Zadeh, L. A. (1975). The concept of a linguistic variable and its application to approximate reasoning—part i. *Information Sciences, 8*(3), 199–249.

Zadeh, L. A. (1996, May). Fuzzy logic = Computing with word. *IEEE Transactions on Fuzzy Systems, 4*(2).

Zadeh, L. A. (2006). Generalized theory of uncertainty (gtu)—principal concepts and ideas. *Computational Statistics & Data Analysis, 51*(1), 15–46. https://doi.org/10.1016/j.csda.2006.04.029.

Zadeh, L. A. (2011). A note on z-numbers. *Information Sciences, 181*(14), 2923–2932. https://doi.org/10.1016/j.ins.2011.02.022.

Zadeh, L. A. (2012). *Computing with words. principal concepts and ideas*. Springer.

Chapter 12
Hands-On Fuzzy Exercises

Abstract This chapter offers several exercises that can be used to test understanding of fuzzy logic and computing with words.

Below are some hands-on exercises on the subject of fuzzy logic and computing with words (e.g., see Bothe, 1995). These exercises are intended to help readers check their understanding and reflect a large part of the topics covered in this textbook.[1]

1. Let the fuzzy sets $\mathbf{A}, \mathbf{B} \in \mathcal{P}(X)$ be given on $X = \{x \in \mathbb{R} \mid 0 \le x \le 2\}$ with the membership functions:

$$\mu_A(x) = \frac{2 - x}{x + 2} \quad \text{and} \quad \mu_B(x) = \frac{(x - 2)^2}{x + 2}.$$

   (a) Calculate the α intersections A_α and B_α for $\alpha = 0.5$.
   (b) Calculate the membership functions for the following expressions:

$$\mathbf{U} = \mathbf{A}^c \cap \mathbf{B}, \quad \mathbf{V} = \mathbf{A} \cup \mathbf{B}^c, \quad \mathbf{W} = \left[\mathbf{A} \cap \mathbf{B}^c\right]^c.$$

   (c) Let a binary operator $\diamond$ be defined by the identity:

$$\mathbf{A} \diamond \mathbf{B} = \left[\mathbf{A}^c \cup \mathbf{B}\right]^c.$$

   Calculate the membership function for the expression:

$$\mathbf{Z} = \mathbf{A}^c \diamond \mathbf{B}.$$

---

[1] Note that the chapter on analog circuit technology is not addressed.

© The Author(s), under exclusive license to Springer Nature Switzerland AG 2026
H.-H. Bothe, E. Portmann, *Computing with Words*, Fuzzy Management Methods,
https://doi.org/10.1007/978-3-032-24117-7_12

2. (a) Show in general that the following relations hold for intersection ($\cap$) and union ($\cup$):

1. $\mathbf{A} \cap (\mathbf{B} \cup \mathbf{A}) = \mathbf{A}$.
   (b) $\forall \mathbf{W} \in \mathcal{P}(\mathbf{X}) : \mathbf{A}, \mathbf{B} \subset \mathbf{W} \Rightarrow (\mathbf{A} \cup \mathbf{B}) \subset \mathbf{W}$.

(b) Let the fuzzy sets $\mathbf{A}, \mathbf{B} \in \mathcal{P}(X)$ be given with $X = [0, 1]$ and the membership functions:

$$\mu_A(x) = \frac{1}{x^2 + 1} \qquad \text{and} \qquad \mu_B(x) = \frac{x}{x^2 + 1}.$$

Calculate the membership functions of the following expressions:

$$U = A \cup B \qquad \text{and} \qquad V = A \cap B.$$

3. Let an alternative T-norm operator on the basic domain X be given by the following formula:

$$\mu_{A \square B}(x) = \frac{\mu_A(x) \cdot \mu_B(x)}{2 - \left[\mu_A(x) + \mu_B(x) - \mu_A(x) \cdot \mu_B(x)\right]}.$$

Calculate the membership function $\mu_{A \blacksquare B}(x)$ of the corresponding S-norm operator.

4. Let $A \blacksquare B$ be an alternative union on the basic domain $X = \mathbb{R}$ given by the following formula:

$$\mu_{A \blacksquare B}(x) = \min\left[1, \left(\mu_A^p(x) + \mu_B^p(x)\right)^{1/p}\right] \qquad \forall x \in X, \ p \geq 1.$$

(a) Show that for $\mathbf{A} = \mathbf{B}$ and $\mu_A(x) = \mu_B(x) \leq 1$, the following applies:

$$\mu_{A \blacksquare B}(x) \geq \mu_A(x) \qquad \forall x \in X.$$

(b) The operator $\blacksquare$ forms an S-norm . For $p = 1$, calculate the membership function $\mu_{A \square B}(x)$ of the corresponding T-norm , and present the result in simple form.

(c) Let the following membership functions be given on $X = [0, 1]$:

$$\mu_A(x) = \frac{1}{x + 1} \qquad \text{and} \qquad \mu_B(x) = \frac{x}{x + 1}.$$

Using $\mu_{Ac, Bc}(x) = 1 - \mu_{A, B}(x)$, calculate the membership functions for:

$$U = A \square B \text{ for } p = 1, \qquad V = A \blacksquare B \text{ for } p = 1.$$

5. Let there be a class of single-digit operators ($\maltese$) for complement formation of fuzzy sets $\mathbf{A} \in \mathcal{P}(X)$ given by the following formula (Yager class):

$$\mu_{A\maltese}(x) = \left[1 - \left(\mu_A(x)\right)^s\right]^{1/s} \qquad \forall x \in X, \ s \in (0, \infty).$$

(a) Show that the following applies:

$$\left(A^{\maltese}\right)^{\maltese} = A \qquad \forall s \in (0, \infty).$$

(b) For the special case $s = 1$, show that the following applies:

$$\mu_{A \cup A\maltese}(x) \geq 0.5, \ and$$

$$\mu_{A \cap A\maltese}(x) \leq 0.5.$$

(c) For the special case $s = 1$, show that the following applies for $\mathbf{A}, \mathbf{B} \in \mathcal{P}(X)$:

$$(\mathbf{A} \cup \mathbf{A}^{\maltese}) \cup (\mathbf{B} \cap \mathbf{B}^{\maltese}) = \mathbf{A} \cup \mathbf{A}^{\maltese}, \ and$$

$$(\mathbf{A} \cup \mathbf{A}^{\maltese}) \cap (\mathbf{B} \cap \mathbf{B}^{\maltese}) = \mathbf{B} \cup \mathbf{B}^{\maltese}.$$

    Take into account the results of 1b.

6. Let a T-norm operator ($\square$) on the basic domain X be given by the following formula:

$$\mu_{A\square B}(x) = \frac{\mu_A(x) \cdot \mu_B(x)}{\max\left[\mu_A(x), \mu_B(x)\right]} \qquad \forall x \in X.$$

(a) Calculate the membership function $\mu_{A\blacksquare B}(x)$ of the corresponding S-norm , and present the result in simple form. To do this, perform a case distinction.
(b) Show that for all fuzzy sets $\mathbf{A}, \mathbf{B}, \mathbf{C} \in \mathcal{P}(\mathbf{X})$, the following applies:

$$(A\square B)\blacksquare(A\square C) = A\square(B\blacksquare C).$$

(c) Specifically, let $\mathbf{A}, \mathbf{B}$, and $\mathbf{C}$ be given on the unit interval $X = [0, 1]$ with the membership functions:

$$\mu_A(x) = x \qquad and \qquad \mu_B(x) = \max[0, 1 - 2x].$$

    Use this to calculate the membership functions of the fuzzy sets $\mathbf{G}$ and $\mathbf{H}$ with:

$$\mathbf{G} = (\mathbf{A}\blacksquare\mathbf{B}^c)^c \qquad and \qquad \mathbf{H} = (\mathbf{A}^c\blacksquare\mathbf{B})\square(\mathbf{A}\blacksquare\mathbf{B})^c.$$

7. (a) Show that for the operators ($\sqcap$) and ($\sqcup$) (bounded product /sum), the second De Morgan's law applies:

$$[\mathbf{A} \sqcap \mathbf{B}]^c = \mathbf{A}^c \sqcup \mathbf{B}^c.$$

(b) Show that $\mathbf{A} \sqcap \mathbf{B} \subseteq \mathbf{A} \cdot \mathbf{B} \;\; \forall x \in X$.

(c) What are the advantages and disadvantages of using the γ operator over the intersection and union operators for quantizing expert statements?

8. (a) Given the two fuzzy sets

$$\mathbf{A_1} = \{(-1; 0.1), \;(0; 1), \;\;\;(1; 0.2)\},$$

$$\mathbf{A_2} = \{(-1; 0.3), \;(0; 0.8), \;(1; 0.4)\},$$

calculate the result $\mathbf{B}$ of the extended multiplication $\mathbf{B} = \mathbf{A_1} \odot \mathbf{A_2}$.

(b) Calculate the expression $\mathbf{D} = (\mathbf{2} \odot (\mathbf{C_1} \ominus \mathbf{C_2})) \oplus \mathbf{C_3}$ for the LR numbers:

$$\mathbf{C_1} = \langle 10; 1; 1 \rangle_{\text{LR}}, \quad \mathbf{C_2} = \langle 5; 2; 3 \rangle_{\text{LR}}, \quad \mathbf{C_3} = \langle 2; 2; 3 \rangle_{\text{LR}}.$$

(c) The derivative $f'(\mathbf{X_0})$ of a crisp function $f(x)$ at a fuzzy point $\mathbf{X_0}$ is defined according to the extension principle for $f'(x) = \mathrm{d}f/\mathrm{d}x$. Calculate $f'(\mathbf{X_0})$ for

$$f(x) = x^3 + 3$$

at the fuzzy point $\mathbf{X_0} = \{(-2; 0.2), (-1; 0.4), (0; 1), (1; 0.5), (2; 0.1)\}$. Note that $\mathbf{X_0}$ is neither positive nor negative.

9. (a) Given the fuzzy sets

$$\mathbf{A_1} = \{(6; 0.1), (12; 1), (18; 0.2)\},$$

$$\mathbf{A_2} = \{(1; 0.3), (2; 1), \;\;\;(3; 0.4)\},$$

calculate the expression

$$\mathbf{B} = \mathbf{2} \odot (\mathbf{A_1} \oplus \mathbf{A_2}).$$

(b) Given the LR numbers $C_1 = \langle 10; 1; 1 \rangle_{\text{LR}}$ and $C_2 = \langle 5; 5; 3 \rangle_{\text{LR}}$ with identical reference functions $L(u) = R(u)$, calculate the expression according to the extension principle:

$$\mathbf{D} = (\mathbf{C_1} \oplus \mathbf{C_2}) \oslash \mathbf{2}.$$

(c) The derivative $f'(\mathbf{C_4})$ of a "crisp" function $f(x)$ at a fuzzy point $\mathbf{C_4}$ is defined according to the extension principle. Calculate $f'(\mathbf{C_4})$ for

$$f(x) = x^3$$

at the fuzzy point $\mathbf{C_4} = \langle -1; 1; 2 \rangle_{\mathrm{LR}}$, $L(u) = R(u) = \max[0, 1 - u]$. Plot the membership functions of $\mathbf{C_4}$ and $f'(\mathbf{C_4})$ in a common diagram.

10. The following membership functions (polygonal curves) of the two fuzzy numbers $\mathbf{A}, \mathbf{B} \in \mathcal{P}(X)$ are given:

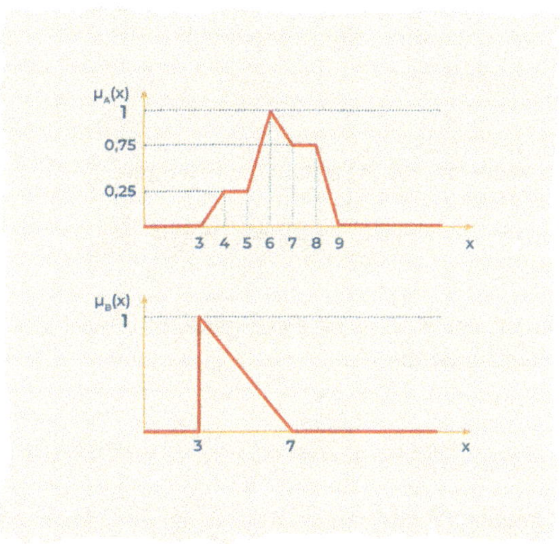

Determine and draw the exact membership functions according to Zadeh's extension principle for the expressions:

(a) $\mathbf{C} = \mathbf{B}^{-1}$ (extended reciprocal formation).
(b) $\mathbf{C}' = \mathbf{A} \oplus \mathbf{B}$.
(c) $\mathbf{C}'' = \mathbf{A} \odot \mathbf{B}$.
(d) $\mathbf{C}''' = \langle 5; 1; 1 \rangle_{\mathrm{LR}} \oplus \mathbf{B}$ with the reference functions

$$L(u) = R(u) = \max[0, 1 - u].$$

(e) Show that for the extended reciprocal formation of a fuzzy number $\mathbf{Z} \in \mathcal{P}(\mathbf{X})$ according to Zadeh's extension principle, the following generally applies:

$$\left(\mathbf{Z}^{-1} \right)^{-1} = \mathbf{Z} \qquad \forall x \in (0, \infty).$$

11. Given the LR numbers

$$\mathbf{C_1} = \langle 5; 1; 1 \rangle_{\mathrm{LR}}, \quad \mathbf{C_2} = \langle 3; 1; 1 \rangle_{\mathrm{LR}}, \quad \mathbf{C_3} = \langle 2; 2; 1 \rangle_{\mathrm{LR}}$$

with the reference functions

$$L(u) = R(u) = \max[0, 1 - u].$$

Calculate:

(a) The expression $\mathbf{U} = 2 \odot (\mathbf{C_1} \oplus \mathbf{C_2}) \ominus \mathbf{C_3}$.
(b) The exact solution for $(\mathbf{C_1})^{-1}$ according to the extension principle.
(c) Let there also be a three-digit function f with:

$$f(x, y, z) = x + 2y + z + 1.$$

Calculate according to the extension principle:

$$\mathbf{Y} = \mathbf{f}(\mathbf{A}, 2 \odot \mathbf{B}, \mathbf{C}).$$

12. (a) Let the fuzzy sets $\mathbf{A}, \mathbf{B} \in \mathcal{P}(X)$ be given on $X = \{-2, -1, 0, 1, 2\}$ with the membership functions

$$\mu_A(x) = \frac{1/2|x|}{x+3} \qquad \text{and} \qquad \mu_B(x) = \frac{1}{x+3},$$

and the function

$$f(x) = x^2 + 1 \qquad \forall x \in X.$$

Calculate $f(A)$ and $f(B)$ using Zadeh's extension principle.
(b) Given the LR numbers

$$\mathbf{C_1} = \langle 6; 1; 1 \rangle_{\text{LR}} \quad \text{and} \quad \mathbf{C_2} = \langle 2; 1; 1 \rangle_{\text{LR}}$$

with the reference functions

$$L(u) = R(u) = \max[0, 1 - u].$$

Calculate the expressions

$$\mathbf{U} = 2 \odot (\mathbf{C_1} \oplus \mathbf{C_2}) \qquad \text{and} \qquad \mathbf{V} = 1/2 \odot (\mathbf{C_1} \ominus \mathbf{C_2}).$$

(c) Using the LR numbers according to b), give an approximate solution for the expression

$$\mathbf{W} = \mathbf{C_1} \odot \mathbf{C_2}$$

that matches the exact solution at $x_0 = 5$, $x_1 = 12$, $x_2 = 21$.

(d) Provide an approximate solution for the expression

$$\mathbf{Q} = \mathbf{W} \oplus \mathbf{C_2}$$

that matches the exact solution to three decimal places.

13. (a) Consider the statement

"Henning drinks (≈ 2) or (≈ 3) liters of milk with dinner"

is given. For quantization, represent the verbal "or" using the bounded sum, the fuzzy numbers as LR numbers with ranges of $a = b = 2$, and the reference functions $L(u) = R(u) = \max[0, 1 - u]$. Calculate and sketch the membership function $\mu_\alpha(x)$ of the term $\alpha = [(\approx 2)$ or $(\approx 3)]$ over the basic domain X with physical unit [liters of milk].

(b) Given the membership function $\mu_{\text{true}}(v)$ of the term *true* for a linguistic variable "truth," calculate the membership function $\mu_{\text{false}}(v)$ of the term *false* from $\mu_{\text{true}}(v)$ by "extending the complement formation" according to the extension principle and sketch it in the adjacent diagram.

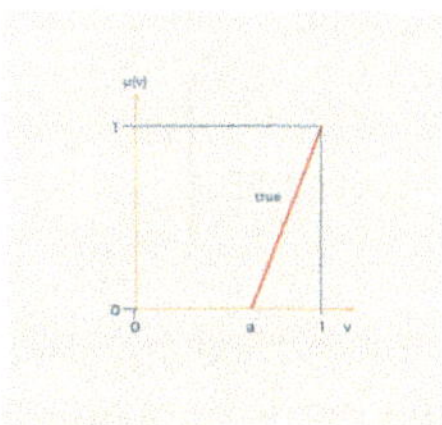

(c) Calculate and sketch the membership function for the expression

(not false) and (not true),

if the verbal *and* is represented by intersection, the verbal *not* by complement formation.

(d) Given the fuzzy relations $\mathbf{R_1}$ and $\mathbf{R_2}$ on $X = [0, 1]$, $Y = [0, 1]$ with the membership functions

$$\mu_{R_1}(x, y) = x^2 y^2 \qquad \text{and} \qquad \mu_{R_2}(x, y) = xy^2,$$

calculate the average $\mu_{R_1 \cap R_2}(x, y)$ and the union $\mu_{R_1 \cup R_2}(x, y)$, and plot the result graphically on $X \times Y$.

14. (a) Let there be a cube with six faces ($\leadsto$ numbers) x, which are color-coded according to the following assignment table:

Face (number)	Color
1, 2	Red
3, 4, 5	Green
6	Blue

An experienced dice roller subjectively assesses the probability $\Pi(\{x\})$ of achieving the individual results $x \in \{1, 2, \ldots, 6\}$ as follows:

x	1	2	3	4	5	6
$\Pi(\{x\})$	0.6	0.2	0.6	0.5	0.8	1

On this basis, assess the probability that the next roll will be "red," "green," "blue," "red," or "green" if exactly one number is rolled with certainty.

(b) The linguistic variable "truth" with a primary term *true* is given. The membership function $\mu_{\text{true}}(v)$ over real numerical truth values $v \in [0, 1]$ is described by:

$$\mu_{\text{true}}(v) = \begin{cases} 0 & 0 \le x < 0.5, \\ 2x - 1 & 0.5 \le x \le 1. \end{cases}$$

The modifier *very* is quantified by concentrating the corresponding membership functions; the logical operators "and," "or," and "not" by the intersection; the union; and the complement of the corresponding fuzzy sets. Calculate the membership functions of the Boolean expressions:

1. False = not true
2. Not (very true)
3. True or (not (very false))

(c) Why does a smaller degree of overlap yield a "better" classifier in terms of the sample size when evaluating fuzzy classifiers?

15. Given the fuzzy sets:

$$\mathbf{A} = \{(1; 0.1), (2; 1), (3; 0.2)\} \in \mathcal{P}(X).$$

$$\mathbf{B} = \{(4; 0.1), (5; 1), (6; 0)\} \in \mathcal{P}(Y).$$

$$\mathbf{C} = \{(3; 0.1), (4; 1)\} \in \mathcal{P}(Z).$$

and the fuzzy relations $\mathbf{R}$, $\mathbf{S}$, and $\mathbf{T}$ with the membership functions

$$\mu_R(x, y) = \max\left[\mu_A(x), \mu_B(y - 2)\right], \quad (x, y) \in X \times Y,$$

$$\mu_S(x, y) = \min\left[\mu_A(x + 2), \mu_B(y)\right], \quad (x, y) \in X \times Y,$$

$$\mu_T(y, z) = \max\left[\mu_B(y), \mu_C(z)\right], \quad (y, z) \in Y \times Z,$$

the complement formation for binary fuzzy relations is defined by:

$$\mu_{Rc}(x, y) = 1 - \mu_R(x, y) \qquad \forall x \in X.$$

(a) Represent **R**, **S**, and **T** in the matrix form, where the x, y, and z values not specified in **A**, **B**, and **C** are assigned the membership value 0.

(b) Represent the following expressions in a matrix form:

$$\mathbf{U} = \mathbf{S}, \quad \mathbf{V} = \mathbf{R} \cap \mathbf{S}, \quad \mathbf{W} = \mathbf{R} \cup \mathbf{S}.$$

(c) Calculate using max-min concatenation:

$$\mathbf{R}' = \mathbf{R} \circ_{MM} \mathbf{T}, \qquad \mathbf{R}'' = \mathbf{S} \circ_{MM} \mathbf{T}.$$

(d) Calculate the largest solution with respect to inclusion $\mathbf{Q}_{max}$ of the relational equation $\mathbf{A} = \mathbf{B} \circ_{MM} Q$.

16. Given the fuzzy relations $\mathbf{R_1}, \mathbf{R_2}, \mathbf{R_3} \in \mathcal{P}(X \times X)$ with the membership functions:

$\mu_{R_1}(x, y):$

$x \backslash y$	1	2	3
1	1	0.4	0
2	0.4	0.8	0.6
3	0	0.4	0.8

$\mu_{R_2}(x, y):$

$x \backslash y$	1	2	3
1	0.1	0.3	0.6
2	0.2	0.6	0.5
3	0.6	1	0.7

$\mu_{R_3}(y, z):$

$y \backslash z$	1	2	3
1	0	0.2	0.4
2	0.1	0.4	0.8
3	1	0.6	0.8

calculate the concatenation results **U**, **V**, and **W** using max-min, max-prod, and max-average concatenation according to:

$$U = \mathbf{R}_1 \circ_{MM} \mathbf{R}_3,$$

$$V = \mathbf{R}_1 \circ_{MP} \mathbf{R}_3,$$

$$W = (\mathbf{R}_1 \cup \mathbf{R}_2) \circ_{MA} \mathbf{R}_3.$$

(a) Show that, in general, the following applies to fuzzy relations $\mathbf{A}, \mathbf{B} \in \mathcal{P}(X \times X)$:

$$\mathbf{A} \circ_{MM} \mathbf{B} \circ_{MM} \neq \mathbf{B} \circ_{MM} \mathbf{A}.$$

17. Given the fuzzy relations $\mathbf{R}_1$, $\mathbf{R}_2$, and $\mathbf{R}_3$ with the membership functions (given as tables):

$\mu_{R_1}(x, y)$:

$x \backslash y$	1	2	3
1	1	0.5	0
2	0.5	1	0.7
3	0	0.7	1

$\mu_{R_2}(y, z)$:

$y \backslash z$	1	2	3
1	0.1	0.2	0.9
2	0.2	0.4	0.5
3	0.5	1	0.1

$\mu_{R_3}(x, y)$:

$x \backslash y$	1	2	3
1	0	0.1	0.2
2	0.3	0.4	0.5
3	1	0.6	0.7

calculate:

(a) The max-min concatenation $\mathbf{R}_1 \circ_{MM} \mathbf{R}_2$,
(b) The max-prod concatenation $\mathbf{R}_1 \circ_{MP} \mathbf{R}_2$,
(c) The union $\mathbf{R}_1 \cup \mathbf{R}_3$ and the intersection $\mathbf{R}_1 \cap \mathbf{R}_3$.
(d) $\mathbf{R}_1$ represents the fuzzy relation "u is approximately equal to v," where the linguistic variables u are defined on X and v on Y. What restriction $\mathbf{B}(y)$ results for v if $u = \mathbf{A} = \{(1; 0.4), (2; 0.8), (3; 1)\} \in \mathcal{P}(X)$ is given?

18. A fuzzy controller is to regulate the "dye addition" w to a liquid depending on its "flow velocity" v. The following rule base is given:

$$\text{IF } v = \text{fast THEN } w = \text{large amount,}$$

$$\text{IF } v = \text{slow THEN } w = \text{small amount.}$$

Let $V = \{1, 2, 3\}$ and $W = \{10, 20, 30\}$ be the basic domains for the linguistic variables v and w. Their terms are defined on these basic domains:

fast	$\rightarrow$	$\mathbf{A}_1 = \{(1; 0), (2; 0.5), (3; 1)\},$
slow	$\rightarrow$	$\mathbf{A}_2 = \{(1; 1), (2; 0.4), (3; 0)\},$
large amount	$\rightarrow$	$\mathbf{B}_1 = \{(10; 0), (20; 0.6), (30; 1)\},$
small amount	$\rightarrow$	$\mathbf{B}_2 = \{(10; 1), (20; 0.3), (30; 0)\}.$

(a) Convert the rule base into a fuzzy relational equation system $\mathbf{B_i} = \mathbf{A_i} \circ_{\text{MM}} R$, and determine the largest solution for the fuzzy relation $\mathbf{R}$ with respect to inclusion.

(b) Determine the fuzzy output value of the rule base when a fuzzy input set $\mathbf{A_3} = \{(1; 0.1), (2; 1), (3; 0)\}$ is present.

(c) Furthermore, let the "temperature" u of the liquid be expressed by the terms hot and cold. For a numerical domain $T = \{100, 200, 300\}$ of the linguistic variable u, these are defined as follows:

hot	$\rightarrow$	$\mathbf{C}_1 = \{(100; 0), (200; 1), (300; 0.6)\},$
cold	$\rightarrow$	$\mathbf{C}_2 = \{(100; 1), (200; 0.2), (300; 0)\}.$

Using the Cartesian product between the corresponding fuzzy sets, define the membership functions in a matrix form for the following states of the liquid:

1. Fast and cold
2. Slow and hot

19. A current-controlled rail vehicle (object) is to be driven to a specific position (target) in the shortest possible time with the aid of a fuzzy controller. The current setting is made depending on the distance between the object and the target on the basis of (fuzzy) expert knowledge. The relationship between current $\mathbf{i}$ [A] and distance $\mathbf{d}$ [m] is given in the following rule base:

$$\text{IF } \mathbf{d} = \text{near} \qquad \text{THEN } \mathbf{i} = \text{small}$$
$$\text{IF } \mathbf{d} = \text{farther} \qquad \text{THEN } \mathbf{i} = \text{medium}$$
$$\text{IF } \mathbf{d} = \text{far} \qquad \text{THEN } \mathbf{i} = \text{large.}$$

The terms of the linguistic variables d and i are quantized as follows using the reference functions $L(u) = R(u) = \max[0, 1 - u]$:

$$
\begin{aligned}
\text{near} &\rightarrow \mathbf{A}_1 = \langle 1; 1; 1 \rangle_{LR}, \\
\text{farther} &\rightarrow \mathbf{A}_2 = \langle 2; 1; 1 \rangle_{LR}, \\
\text{far} &\rightarrow \mathbf{A}_3 = \langle 3; 1; 1 \rangle_{LR},
\end{aligned}
$$

$$
\begin{aligned}
\text{small} &\rightarrow \mathbf{B}_1 = \langle 10; 0; 0 \rangle_{LR}, \\
\text{medium} &\rightarrow \mathbf{B}_2 = \langle 20; 0; 0 \rangle_{LR}, \\
\text{large} &\rightarrow \mathbf{B}_3 = \langle 30; 0; 0 \rangle_{LR}.
\end{aligned}
$$

The fuzzy sets $\mathbf{B}_1$, $\mathbf{B}_2$, and $\mathbf{B}_3$ should represent singletons:

(a) Represent the functioning of the rule base graphically using a diagram.

(b) Using the max-min inference method, determine the fuzzy output values $\mathbf{i}_1$ and $\mathbf{i}_2$ of the rule base for the crisp input values $\mathbf{d}_1 = 1.5$ and $\mathbf{d}_2 = 2.5$.

(c) Using the max-min inference method, determine the fuzzy output value $\mathbf{i}_3$ of the rule base for a fuzzy input value $\mathbf{d}_3 = \langle 1.5; 0.5; 0.5 \rangle_{LR}$.

(d) Using the max-min inference method, determine the fuzzy output value $\mathbf{i}_4$ of the rule base for a fuzzy input value $\mathbf{d}_4 = \langle 1.8; 0.5; 0.5 \rangle_{LR}$.

(e) What is the (crisp) setpoint $\mathbf{i}_4$ resulting from after defuzzification using the max method?

20. A balance beam is to be kept in a horizontal position with the help of a fuzzy controller. For this purpose, an electric motor can be moved back and forth on the beam by means of voltage control. This is done depending on the angle of rotation a and the distance d : motor $\leftrightarrow$ pivot point on the basis of fuzzy expert knowledge.

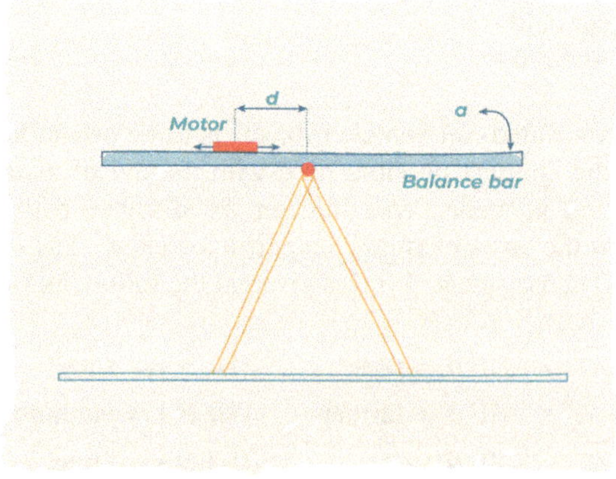

The relationship between the voltage **u** [V] to be set, the measured distance **d** [cm], and the measured angle of rotation **a** [°] is given in the following rule base:

$$\begin{array}{ll}
\text{IF } \mathbf{d} = \text{positive AND } \mathbf{a} = \text{positive} & \text{THEN } \mathbf{u} = \text{ng} \\
\text{IF } \mathbf{d} = \text{positive AND } \mathbf{a} = \text{negative} & \text{THEN } \mathbf{u} = \text{nk} \\
\text{IF } \mathbf{d} = \text{negative AND } \mathbf{a} = \text{positive} & \text{THEN } \mathbf{u} = \text{pk} \\
\text{IF } \mathbf{d} = \text{negative AND } \mathbf{a} = \text{negative} & \text{THEN } \mathbf{u} = \text{pg.}
\end{array}$$

The terms of the linguistic variables are abbreviated using reference functions $L(u) = R(u) = \max[0, 1 - u]$ in LR notation:

$$\begin{array}{lll}
\text{positive} & \rightarrow & \mathbf{A}_1 = \langle 1; 2; 2 \rangle_{\mathrm{LR}}, \\
\text{negative} & \rightarrow & \mathbf{A}_2 = \langle -1; 2; 2 \rangle_{\mathrm{LR}},
\end{array}$$

$$\begin{array}{lll}
\text{ng} & \rightarrow & \mathbf{B}_1 = \langle -20; 0; 0 \rangle_{\mathrm{LR}}, \\
\text{nk} & \rightarrow & \mathbf{B}_2 = \langle -10; 0; 0 \rangle_{\mathrm{LR}}, \\
\text{pk} & \rightarrow & \mathbf{B}_3 = \langle 10; 0; 0 \rangle_{\mathrm{LR}}, \\
\text{pg} & \rightarrow & \mathbf{B}_4 = \langle 20; 0; 0 \rangle_{\mathrm{LR}}.
\end{array}$$

The fuzzy sets $\mathbf{B}_1$, $\mathbf{B}_2$, $\mathbf{B}_3$, and $\mathbf{B}_4$ should represent singletons:

(a) Represent the functioning of the inference engine graphically using an overview diagram.
(b) Calculate the output value $\mathbf{u}_1$ using the max-min inference method for the (crisp) input values $d_1 = 0.5$ and $a_1 = 0.5$, and sketch the calculation process in the diagram according to a).
(c) Which (crisp) control value u_s results from b) after defuzzification using the centroid method?
(d) What are the advantages and disadvantages of rules in predictive logic compared to those in the rule base specified here, and why?

21. The transfer behavior of a complex system G can be described with the help of fuzzy expert knowledge as shown in the following block diagram.

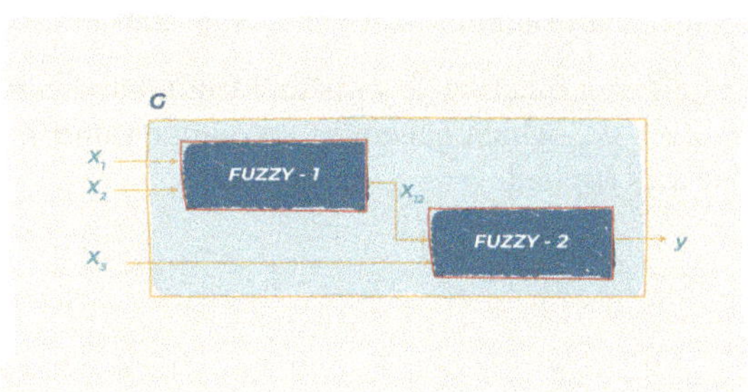

The sharp input values $x_1, x_2, x_3 \in [-3, 3]$ are processed by two fuzzy inference engines, FUZZY-1 and FUZZY-2, using the max-min inference method. FUZZY-1 delivers the fuzzy output result $\mathbf{x_{12}}$; this, together with x_3, serves as the input variable for FUZZY-2. The variability range of the output quantity $\mathbf{y}$ lies in the interval $[-0.5, 0.5]$. To describe the transfer behavior of FUZZY-1 and FUZZY-2, the following terms and associated fuzzy sets with $L(u) = R(u) = \max[0, 1 - u]$ are first defined:

$$\begin{array}{rcl}
\text{ne} & \rightarrow & \mathbf{A} = \langle -1; 2; 2 \rangle_{\mathrm{LR}}, \\[4pt]
\text{po} & \rightarrow & \mathbf{B} = \langle 1; 2; 2 \rangle_{\mathrm{LR}},
\end{array}$$

$$\begin{array}{rcl}
\text{kl} & \rightarrow & \mathbf{S_1} = \langle -0.5; 0; 0 \rangle_{\mathrm{LR}}, \\[4pt]
\text{mi} & \rightarrow & \mathbf{S_2} = \langle 0; 0; 0 \rangle_{\mathrm{LR}}, \\[4pt]
\text{gr} & \rightarrow & \mathbf{S_3} = \langle 0.5; 0; 0 \rangle_{\mathrm{LR}}.
\end{array}$$

(a) The rule base of FUZZY-1 is described using linguistic variables LV_1, LV_2, LV_{12} via x_1, x_2, x_{12} and the terms specified above and reads:

IF $LV_1 = ne$ AND $LV_2 = po$ THEN $LV_{12} = mi$
IF $LV_1 = po$ AND $LV_2 = ne$ THEN $LV_{12} = kl.$

Determine and plot the fuzzy output value $\mathbf{x_{12}}$ for the input value vector:

$$(x_1, x_2, x_3)^\top = (0, -1/2, 1/2)^\top.$$

(b) The rule base of FUZZY-2 is described using linguistic variables LV_{12}, LV_3, LV_y via x_{12}, x_3, y and the terms specified above and reads:

IF $LV_{12} = ne$ AND $LV_3 = po$ THEN $LV_y = gr$
IF $LV_{12} = po$ AND $LV_3 = ne$ THEN $LV_y = kl.$

Determine and plot the fuzzy output value y for the input value vector:

$$(x_1, x_2, x_3)^\top = (0, -\tfrac{1}{2}, \tfrac{1}{2})^\top.$$

(c) Defuzzify the fuzzy output result via $\mathbf{y}$ using the centroid method.

22. The course of a given function $f(x)$ is to be simulated using a fuzzy inference machine "FUZZY-f," which generates an output value $f_{\mathrm{out}}(x) \approx f(x)$ when an input value x is applied.

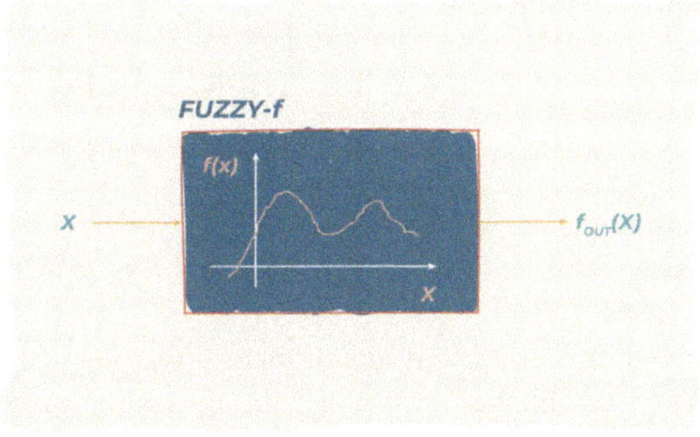

Given the functions

$$f_0(x) = x^2 \qquad\qquad \text{for } x \in [0, 10],$$

$$f_1(x) = x^3 - 2x^2 + x \qquad \text{for } x \in [-2, 2],$$

$$f_2(x) = \sin^2(x) \qquad\qquad \text{for } x \in [0, 2\pi],$$

$$f_3(x) = 1/x\ \sin(x) \qquad\quad \text{for } x \in [0, 2\pi].$$

(a) Sketch the function curves of $f_0(x)$, $f_1(x)$, $f_2(x)$, and $f_3(x)$ in the specified definition ranges.

(b) Set up a rule base for approximating the function curves of $f_0(x)$, $f_1(x)$, $f_2(x)$, and $f_3(x)$. Start from the model of overlapping fuzzy points $\mathbf{Q_i}$ to visualize the production rules "IF $u = \mathbf{A_i}$ THEN $v = \mathbf{B_{i'}}$, $\mathbf{A_i} \in \mathcal{P}(X)$, $\mathbf{B_i} \in \mathcal{P}(Y)$." Place the fuzzy points at characteristic points of the function to be approximated according to the following principle:

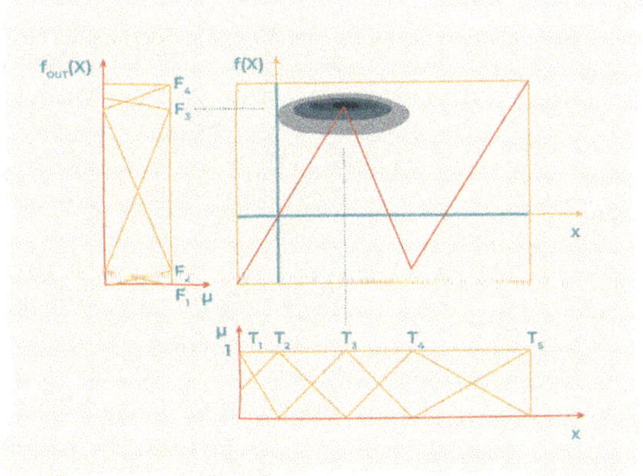

Example of a rule base:

$$
\begin{aligned}
&\text{If } u = \mathbf{T_1} \quad \text{THEN} \quad v = \mathbf{F_1} \\
&\text{If } u = \mathbf{T_3} \quad \text{THEN} \quad v = \mathbf{F_3} \\
&\text{If } u = \mathbf{T_4} \quad \text{THEN} \quad v = \mathbf{F_2} \\
&\text{If } u = \mathbf{T_5} \quad \text{THEN} \quad v = \mathbf{F_4}
\end{aligned}
$$

(c) Sketch the basic curves of the transfer functions of the fuzzy systems "FUZZY-f_i" established in **b** using the maximum method for defuzzification.

(d) How does the approximation error $\varepsilon = \int \left[f_{\text{out}}(x) - f(x) \right]^2 dx$ depend on the choice of membership functions and the position of the fuzzy points?

23. Computing with words allows us to construct mathematical solutions of computational problems which are stated in natural language. Solve the following natural-language problems:

(a) Most Swiss are tall. What is the average height of the Swiss?

(b) Probably your neighbor is tall. What is the probability that he is short? What is the probability that he is very short? What is the probability that he is not very tall?

(c) Usually, most Swiss flights from Zurich leave on time. You are scheduled to take a Swiss flight from Zurich. What is the probability that your flight will be delayed?

Reference

Bothe, H.-H. (1995). *Bewertung mit unscharfen Mengen*. Technische Hochschule Ilmenau.

Index

© The Editor(s) (if applicable) and The Author(s), under exclusive license to   207
Springer Nature Switzerland AG 2026
H.-H. Bothe, E. Portmann, *Computing with Words*, Fuzzy Management Methods,
https://doi.org/10.1007/978-3-032-24117-7

9 783032 241160